£4.99

Ion Plating Technology

Ion Plating Technology

Developments and Applications

N. A. G. AHMED

IonCoat Consultancy Services,
Manchester, UK

JOHN WILEY & SONS
Chichester · New York · Brisbane · Toronto · Singapore

Library of Congress Cataloging-in-Publication Data:

Ahmed, N. A. (Nadir A.)
Ion plating technology.

Bibliography: p.
1. Ion plating. I. Title.
TS695.2.A36 1987 671.7'35 86-32476
ISBN 0 471 91473 8

British Library Cataloguing in Publication Data:

Ahmed, N. A.
Ion plating technology.
1. Ion plating
I. Title
671.7'35 TS695.2

ISBN 0 471 91473 8

Phototypeset by Dobbie Typesetting Service, Plymouth, Devon
Printed and bound in Great Britain

DEDICATION

To R.M.K., my Wife, Son and Daughter with Appreciation

Preface

Ion plating is a hybrid vacuum coating process that combines the benefits of vacuum evaporation and sputtering. The term ion plating is generally applied to high energy plasma deposition methods in which the surface to be coated is subjected to a small flux of high energy ions and a much larger number of energetic neutrals before and during the deposition of the coating. The continuous bombardment of the substrate by these energetic ions and atoms of both the coating material and the support gas results in a wide variety of effects. The four most important physical processes occurring during the ion plating process are:

1. Substrate cleaning due to the desorption of absorbed impurities from the substrate surface as a result of ion and atom bombardment.
2. Ion penetration and entrapment in the substrate and coating.
3. Ion-induced sputtering of substrate and coating atoms.
4. Recoil displacement of substrate and coating atoms leading to their intermixing.

The physical effects of the action of such processes consequently result in an excellent adhesion of the coating to the substrate, denser coating with little or no porosity and good chemical composition. For the last 22 years, the ion plating process has been developed from a pure laboratory research technique to a potential industrial coating process where it has proliferated into many areas in which other coating processes have not been so successful.

I would like to thank many colleagues who have stimulated and encouraged me in writing this book. I would particularly like to thank Dr J. S. Colligon, Professor D. G. Armour, Dr D. Arnell and Dr C. W. Smith for their important comments, valuable suggestions and encouragement in this venture. I would also like to thank authors, publishers, companies and institutions who kindly granted me permission to reproduce some of their results and experimental techniques in this book. Finally, I would like to thank my wife, Wedad, for her support and patience during the preparation of this book.

Nadir A. G. Ahmed

Contents

Chapter 1
Introduction

In many applications, surface coatings are used for a variety of reasons including the requirement for better physical and chemical properties to improve the surface of bulk materials, economics, materials conservation and design flexibility, since surface properties can be separated from the bulk properties. The applications of such coatings include optical coatings for lenses and mirrors, metallic coatings for electrical contacts, refractory coatings to increase hardness and decrease wear, and protective coatings for corrosion and oxidation resistance. The general requirements for these coatings are: excellent adhesion to the substrate, dense and pore-free structure, good uniformity and smooth finish. Surface coatings can be deposited by various methods including physical (PVD) and chemical vapour deposition (CVD) and electroplating. Of these methods, physical vapour deposition has received the greatest interest due to its flexibility in terms of depositing any material, whether single or alloy, adhesion, throwing power (i.e. ability to coat complex parts), cost and ecological considerations. The physical vapour deposition technique (PVD) can be classified as evaporation deposition using resistance induction, arc, plasma, electron beam or thermionic electron beam evaporation sources, or sputter deposition using diode, triode, or magnetron (planar, cylindrical or post cathode) sputter sources, or a combination of these two processes. Ion plating falls in the third category and is considered as a hybrid process since the vapour can be generated either by evaporation or sputtering. Although ion plating is a relatively new coating process as compared with evaporation or sputtering, its concept goes back as early as 1938 when Berghaus (1938) described a vacuum coating technique in which vapour was deposited on a substrate in a glow discharge. The coating was claimed to have a perfect structure and adhering strength even for thicker layers. Mattox (1963) introduced a similar technique in which he used a glow discharge to ionize the atoms of gas and coating material and accelerated them through a suitable potential towards the substrate which was the cathode of the discharge. The process was named 'Ion Plating' and it was claimed that it produced films with excellent adhesion, even in the cases where the film and substrate material were mutually incompatible (Mattox, 1964). This work has stimulated great interest and since then vast numbers of papers have been published.

The term 'Ion Plating' is now generally applied to high energy plasma deposition methods in which the surface to be coated is subjected to a small flux of ions having energies up to (1–2) keV and a much larger number of energetic neutrals before and during the deposition of the coating (Spalvins *et al.*, 1966; Chambers and Carmichael, 1971; Mattox, 1971; Ahmed, 1974; NASA, 1972). The physical effects of bombarding the film–substrate interface with energetic ions and neutrals permit an excellent adhesion for a wide range of film substrate combinations, good coverage, and produces coatings with little or no porosity, a property which is essential for the production of corrosion protective coatings (Burns and Bradly, 1967). Although the name of 'Ion Plating' seems to have gained general acceptance, Miles (1968) has suggested that 'Ion Plating' is an erroneous term since ionized metal species are only a small fraction of the condensed particles that form the coating. For that reason other names are frequently used including 'Ion Vapour Deposition' or 'Ion Assisted Coatings' to indicate the same process. Furthermore, variations of the ion plating techniques have given 'Chemical Ion Plating', 'Alternating Ion Plating' and others which relate to specific environments or techniques (Mattox, 1982a).

During the last two decades the activity and interest in ion plating has been developed on two levels, the fundamental or laboratory level and the industrial or technological level. On the fundamental level, there has been a noticeable trend towards a better understanding of the physics of the process which generally will lead to a better control of the coating properties and the discovery of the optimum plating conditions (Ahmed, 1974). On the industrial level, ion plating has proliferated into many areas in which other coating techniques have not been successful. For instance, ion plating technology is now well established in the aerospace, aircraft, and tool industries for the application of tribological coatings to reduce friction and wear and protective coatings to resist corrosion and oxidation. The technique is currently used on a production scale to aluminize aircraft components for corrosion protection or to deposit hard ceramic coatings such as titanium nitride on cutting tools for wear resistance (Ind. Finish., 1978).

The purpose of this book is to give the reader an introduction to the physics of the ion plating process, to classify the numerous ion plating techniques, to describe the characteristics of the ion plated coatings and to describe potential applications of ion plating in different fields. The contents of this book are mainly intended for postgraduate students in the fields of physics, engineering and metallurgy, as well as researchers and engineers working in metallurgy or the coating industry. It is expected that these readers will have a basic knowledge of vacuum physics and deposition methods. However, because of the widely varying backgrounds of the anticipated readers, certain sections have been included to introduce the reader to the fundamental aspects of gas discharges and sputtering before discussing in depth the different processes occurring during ion plating. Also, an attempt has been made to report up-to-date applications of ion plating. However, because of the vast numbers of new publications and because of the other new applications of the modified versions of ion plating such as ion assisted and plasma deposition techniques, this attempt has proved to be difficult. Therefore, only the most common industrial applications of ion plating have been reported, with reference to a number of newly born applications which are still on a laboratory scale. It is hoped that this book will help the reader to understand the physics of the ion plating process and to select one of the many forms of ion plating which will meet his requirements. Furthermore, it is also hoped that on the basis of such understanding, the reader will be able to contribute further to the expanding field of ion plating.

Chapter 2
The Characteristics of the Ion Plating Process

Before discussing the ion plating technique in more detail, an attempt will be made to review the physics of gas discharges and sputtering, thus giving a general background to the physics of ion plating.

2.1 THE GAS DISCHARGE

It has been known for a long time that if a d.c. voltage is applied between two electrodes in a vacuum chamber containing a gas at a pressure greater than about 0.1 Pa, then a discharge can be initiated between the electrodes (Parker, 1950). The primary source of ionization of the discharge gas is the electron–atom collision (Mattox, 1971), where collision between electrons and gas atoms results in the ionization of gas atoms and the emission of more electrons:

$$e^- + G^0 \rightarrow G^+ + 2e^- \qquad (2.1)$$

where G^0 is a ground state gas atom and G^+ is a singly charged gas ion.

The current–voltage characteristics of such a d.c. gas discharge usually have three distinct regions as shown in Figure 1, (Howatson, 1975). In the first region and at low voltage, little current is drawn because only a few ionized particles exist which can contribute to the current. As the voltage increases, the charged particles gain sufficient energy to produce additional charged particles by impact ionization. This leads to a linear growth of the discharge current, the voltage thereby remaining constant, limited by the high output impedance of the voltage supply. This region of the discharge is usually called a 'Townsend' discharge. As the voltage increases further, an avalanche process takes place due to the ion bombardment of the cathode and the release of secondary electrons. Such electron emission is usually observed following a bombardment by ions, electrons, photons and neutrals, but each will have a different emission coefficient and different energy dependence (Chapman, 1980). The secondary electrons are then accelerated in the field of the cathode and due to collisions with the residual gas atoms and molecules produce new ions as shown in equation (2.1). The ions are again accelerated back to the cathode and produce new secondary electrons. This process continues until a self-sustaining discharge is achieved. In this case, the gas starts to glow, the voltage falls and the current rises abruptly. This region is called the 'Normal' discharge region where at a constant

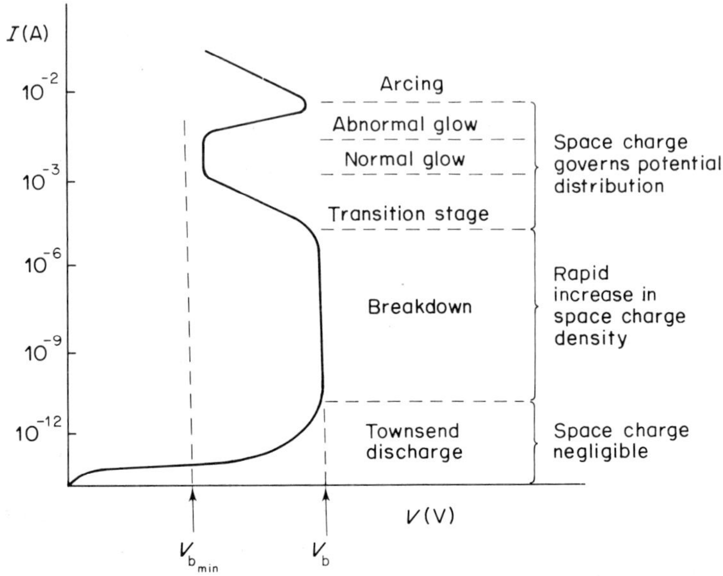

Figure 1. Overall characteristics of a gaseous discharge

voltage, the current is a function of gas pressure and the cathode current density is constant over the cathode spot. As the voltage increases further, a uniform current intensity distribution can be observed and the current density becomes a function of the voltage at a constant pressure (Parker, 1950; Howatson, 1975). This region is called the 'Abnormal Glow Discharge' and is the region where sputtering and ion plating are carried out. When viewing the abnormal glow discharge, a number of dark and bright zones can be distinguished and the most prominent ones are shown in Figure 2. The light region near the cathode is called the cathode glow. The colour of the glow is characteristic of the cathode material. This is due to the decay of atoms sputtered in excited states. The contribution of charge exchange processes which are indicated in equation (2.2) is not very significant since, even though these processes are occurring, they do not involve excited states to a large enough extent to account for all emission observed.

$$\underset{\text{Fast}}{G_1^+} + \underset{\text{Slow}}{G_2^0} \rightarrow \underset{\text{Slow}}{G_2^+} + \underset{\text{Fast}}{G_1^0} \qquad (2.2)$$

A dark space arises in front of the cathode glow and in this zone the secondary electrons are accelerated away from the cathode. The width of the dark space (d) is inversely proportional to the gas pressure (P) so that:

$$Pd \simeq \text{constant}$$

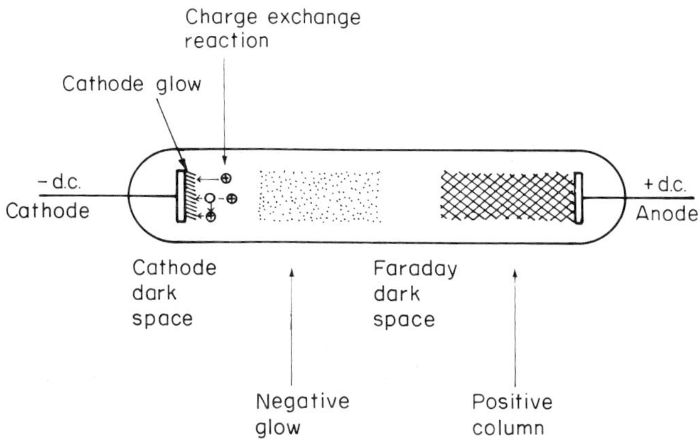

Figure 2. Light and dark zones in an abnormal glow discharge

The width of the dark space represents the number of electron mean ionization path lengths required for secondary electrons to produce enough ions to maintain the discharge and is determined by pressure, type and temperature of the gas. For example, at 1 Pa argon discharge pressure, the cathode dark space is about 2 cm wide (Chapman, 1980). Furthermore, in the dark space the positive ions are accelerated to the cathode. The velocity of the ions is very much smaller than that of the electrons and the ion density is high. Thus a high space charge build-up appears as shown in Figure 3. At the end of the dark space another light zone arises and is called the 'Negative Glow'. In this zone the first ionizing collisions of electrons with neutral gas atoms take place and thus the production of ions occurs. However, in the dark space electrons do collide with gas atoms but have insufficient energy to cause ionization. The Faraday dark space and the positive column adjoining the region of the negative glow present almost field-free spaces, as can be derived from the potential curve and from the electric connection to the anode.

2.2 SPUTTERING

Sputtering is generally defined as a process of dislodging and ejecting material from the surface of a solid due to the momentum exchange associated with surface bombardment by energetic particles (Holland, 1974a). The coating processes which exploit sputtering may take a number of different forms. One of these, the diode arrangement, consists of two electrodes installed inside a vacuum chamber which is usually evacuated to a pressure from 5×10^{-2} to 5×10^{-5} Pa. One of the electrodes is the source of coating material and is called the 'target' and is attached to a high negative potential from 10 to 5000 V. The other electrode is the substrate holder and is situated opposite the target. After evacuating the chamber to a suitable pressure it is backfilled with an inert

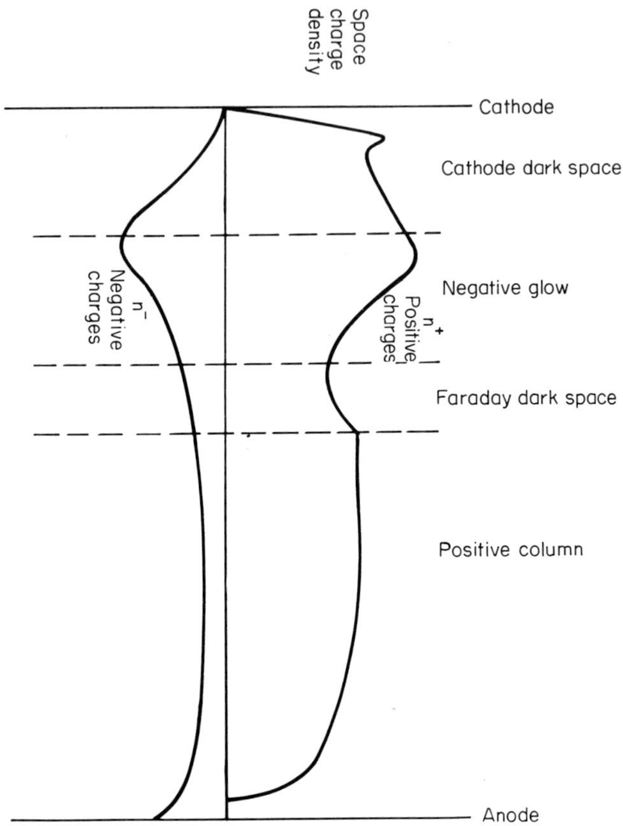

Figure 3. Schematic representation of the positive space charge density in a gaseous discharge

gas (normally argon) to a pressure of 0.1–10 Pa. The negative potential on the target 'ignites' the discharge which is then maintained by electrons produced at the cathode by positive ion bombardment. In this way, the ionization of the working gas is produced in the region adjacent to the target (Figure 4). Such a low pressure electric discharge is called abnormal glow discharge and the ionized gas is called plasma. As a result of ionization, energetic particles including positive argon ions bombard the target leading to the ejection of atoms from the surface by momentum exchange. These atoms eventually condense on the substrate surface to form a continuous film. However, in order to commence sputtering the ions incident on the target must have a minimum energy (threshold energy) which exceeds the binding (sublimation) energy of the target atoms (usually 2–10 eV). Therefore, ions bombarding the target at certain energies transfer their energy to the target atoms through a series of multiple collisions generated by the primary collision at the surface. Target atoms are ejected either

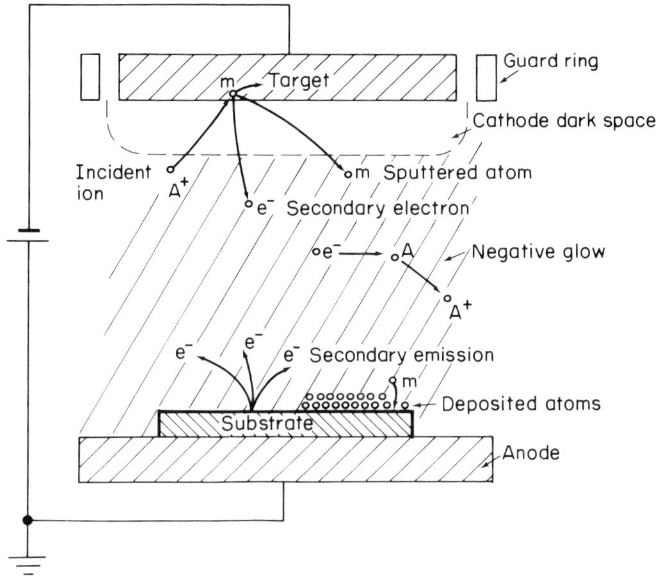

Figure 4. Schematic representation of a simple d.c. diode sputtering process

by a direct collision between the impinging ion and the surface atom or by penetration of the ion into the solid producing internal collisions with recoils which may be sputtered from the surface if they receive energies greater than the surface binding energy (U_o). The latter process, which involves a cascade of moving target atoms, is known as a collision cascade and is responsible for the majority of sputtered particles. The number of target atoms ejected per incident ion is defined as the sputtering yield (Y). Over many years, there have been considerable attempts to predict the sputtering yield theoretically in terms of projectile and target parameters, and even more experimental studies for comparison. However, the most widely accepted current theory of sputtering which fits experimental data extremely well in the linear cascade region is that due to Sigmund (1969). Sigmund's prediction for the sputtering yield for projectiles incident normally on a target (which will be most relevant to ion plating situation) (Carter and Armour, 1981) is:

$$Y(E) \simeq \frac{0.042\ \alpha\ Sn\ (E)}{U_0} \tag{2.3}$$

where $Sn(E)$ is the energy loss per unit distance by a projectile as it enters the solid, U_0 is the binding energy of surface atoms and α is a constant rather dependent upon the ratio M_2/M_1 of target-to-projectile atomic masses, varying from about 0.15 for $M_2/M_1 = 0.1$ to 1.5 for $M_2/M_1 = 10$, and weakly dependent upon other parameters. However, this prediction is inaccurate in sputtering by low energy projectiles ($\leqslant 1$ keV). This led Sigmund (1969) to propose a low energy modification:

$$Y(E) \simeq \frac{3}{4\pi^2} \frac{\gamma E}{u_0} \tag{2.4}$$

where $\gamma = 4\, M_1M_2/(M_1+M_2)$ is the energy transfer coefficient and is valid for $E \gg U_0$. *E is the energy of the incident ion.*

This expression for Y predicts that the yield will increase linearly with E. In practice, this seems to be satisfied up to 1 KeV, above which Y becomes relatively constant. At higher energies, Y decreases as ion implantation becomes dominant.

In general terms, the sputtering yield depends on the relative masses of projectiles and target atoms, energy of incident particles, structure and crystallographic orientation of the target, target material, angle of incidence of the bombarding particle, surface morphology and gas pressure. It is independent of target temperature or whether the bombarding species are ionized or not. The yield tends to be greatest when the mass of the bombarding particles is not of the same order of magnitude or larger than that of the target atoms. For this reason, argon is generally used because of its mass compatibility with materials and its non-reactivity with the target or substrate, as well as its low cost. However, krypton or xenon are sometimes used in certain applications. For example, in the case of gold, the sputtering yield for Xe^+ ions is about three times higher than with Ar^+ ions.

Another important factor which depends on the sputtering yield is the erosion rate of target (R). This expression is used in estimating the amount of material removed from the target during sputter cleaning and bias sputtering. The erosion rate is given by:

$$R = 62.3 \frac{J\, YM_a}{\varrho} \quad \text{Å min}^{-1} \tag{2.5}$$

where J is the ion current density in mA cm^{-2}, Y is the sputtering yield in atoms per ion, M_a is the atomic weight of the target in grams and ϱ is the density in g cm^{-3} of the target material.

In view of the importance of the processes occurring in the glow discharge on the sputtering mechanism, these processes will be discussed in detail in the following sections.

2.3 PROCESSES OCCURRING NEAR THE CATHODE

When an energetic ion or neutral atom hits the surface of the sputtering target (cathode) various complicated processes may occur simultaneously, in addition to production of the secondary electrons which serve to maintain the discharge. The high energy ions/atoms that impinge on the cathode may sputter either cathode(s) or contaminant atoms from the surface, or be reflected as high energy ions or neutrals. The neutral particles leaving the cathode are often found to be in an excited state (G^*). The impinging particles also sputter film atoms (M) in the case of bias sputtering or ion plating. In this case it is necessary for the deposition rate to exceed the sputtering rate for a film to grow. However, the sputtered surface atoms (film, substrate or contaminant) may be either:

(1) scattered back to the cathode; or
(2) ionized by electron atom transition during collision between an excited metastable atom (G^*) and a neutral atom (S), i.e.

$$S^0 + G^* \rightarrow S^+ + G^0 + e^- \text{ (Penning Ionization)} \tag{2.6}$$

and accelerated back to the cathode or lost from the cathode region; or
(3) transported as high energy neutrals of the target material to the same surface.

According to Mattox (1980), Penning ionization is probably the most important ionization process for sputtered cathode atoms. The return of sputtered atoms to the cathode surface is a function of many discharge parameters, as well as the system throughput. This return is important in sputter cleaning, generation of surface features (such as cones and whiskers), surface coverage (ion plating) and the generation of the pseudo-diffusion type of interface (Mattox, 1971). It has been estimated that at 10 Pa argon gas pressure, 90% of the sputtered atoms are scattered back to the cathode (Mattox, 1982a). Furthermore, the injected atoms act as foreign particles in the gas discharge and can cause changes in the discharge parameters. Beside the above processes, other processes may take place at the cathode (target) and at the substrate in the case of bias sputtering and ion plating. Such processes are temperature increase, emission of radiation and chemical reactions and dissociation. Temperature increase occurs since 90% of ion energy on the target is lost as heat in the solid material and only 5% is passed onto the secondary particles. The emitted radiation from the target surface consists mainly of photons in the ultraviolet and visible spectral range. The radiation results from particles and ions excited by resonance or Auger electron transition (Mattox, 1982b). Hence, it is a characteristic of the target material and the entering ions and can be used for process analysis. The u.v. radiation may damage photoconductors or special plastics. Chemical reactions and dissociation are caused by electron bombardment of atoms and molecules, thereby leaving a variety of active species and radicals having very different chemical activities from those of the parent gas. Such excitation and dissociation processes are important in plasma chemistry and form the basis for plasma-assisted etching, reactive ion plating, plasma-assisted chemical vapour deposition (CVD) and plasma polymerization (Blocher, 1982).

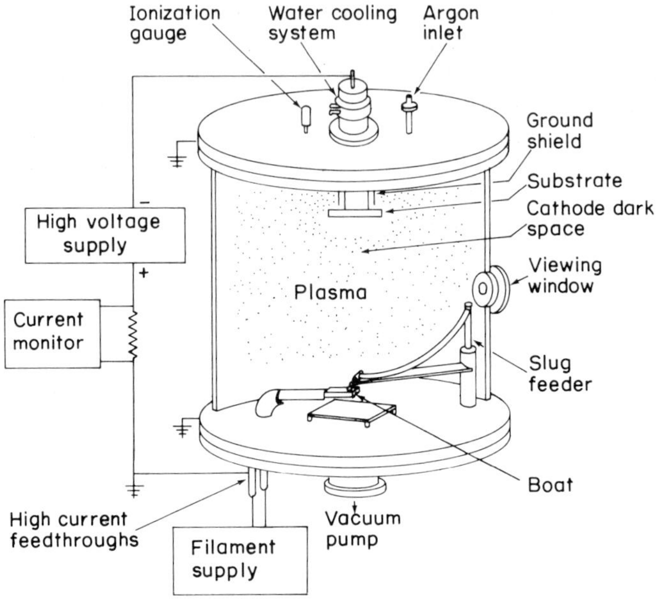

Figure 5. Schematic diagram of a basic ion plating set-up

2.4 THE ION PLATING PROCESS

The ion plating process is generally carried out in an inert gas discharge system similar to that used in sputter deposition, except that the substrate is made as the sputtering cathode (Mattox, 1966). Figure 5 shows a schematic diagram of a basic ion plating set-up. The substrate is continuously bombarded by the energetic positive ions and neutrals of the inert gas and subjected to the action of the ultraviolet radiation emitted from the positive column of the abnormal glow discharge (Mattox, 1974a). The ionic bombardment sputter cleans and modifies the substrate surface on an atomic scale prior to the deposition of the film material from an evaporation source. As described in Section 2.2, the sputter cleaning depends on many factors such as sputtering yield of the target (substrate) material, the current density and the ion energy. Therefore, while the actual cleaning time for copper may take only 15 minutes, chromium may require up to 60 minutes (Teer and Halling, 1978). After cleaning is complete, the coating material is evaporated without interrupting the ion bombardment and for a film to form, it is necessary for the deposition rate to exceed the re-sputtering rate. The gas pressure used during the cleaning stage and the subsequent evaporation stage is much higher than that used in vacuum evaporation. The ion plating process is usually carried out at argon gas pressure in the range of 0.1 – 1 Pa (Mattox, 1974b). This pressure is needed to create conditions for an abnormal glow discharge to operate and hence produces the necessary ions required to maintain the glow discharge as well as scattering the metal vapour in all directions through numerous collisions to improve the overall coating of the substrate in the process. Such scattering of metal vapour is called the 'throwing power'. Ionization of some of the evaporated metal atoms occurs as a result of collisions with electrons, atoms and ions in the plasma. As a result of the bombardment of substrate and coating, different substrate surface, nucleation and growth conditions exist resulting in changes in flux composition and other properties (Schiller *et al.*, 1979).

Although the superior mechanical properties of ion plated coatings such as adhesion, friction, hardness and wear have been appreciated and utilized for many years, the process is still not fully understood. In particular, little is known about the basic mechanisms occurring during film deposition and the energy spread of the ions reaching the substrate prior to and during the evaporation process. The following review presents the available background information on ion plating conditions and cites evidence indicating which of these are important in determining specific film properties.

2.5 NUMBER OF IONIZED PARTICLES

The number of ionized particles in a glow discharge depends on the balance between the rate at which ionization is produced by the energetic electrons and the rate at which particles are lost by volume recombinations and by passage to the walls of the apparatus (Thornton and Hoffman, 1981). Therefore, the rate of ionization depends on the type of gas (through the ionization cross-section), the gas pressure (through the particle density) and the electric field strength (through the electron velocity). Wall losses generally dominate over volume recombination. Therefore, the formation of a sustaining plasma discharge in a given apparatus depends on the gas pressure, electric field strength and on the surface-to-volume ratio of the plasma. When the atoms of the coating material are injected into the glow discharge during the ion plating process, a small fraction is ionized primarily by an electron–atom collision or by a metastable atom–atom collision (Penning ionization). The low degree of ionization is primarily attributed to the small ionization cross-section and wall losses. Many attempts have been made by different workers to estimate the number of ionized particles during the ion plating process (Teer *et al.*, 1976; Teer and Halling, 1978; Aisenberg, 1972). Controversial figures have been reported ranging from 1% up to 30%. The figure of 30% was based on the assumption that film thickness on the back surface of a substrate is 30% lower than that on the front surface and the reduction is caused by the fact that 30% of the metal ions being ionized will follow the field lines and deposit on the back surface. However, such an assumption greatly overestimates the number of ionized particles because the major factor causing back surface coating is the scattering of metal atoms by the inert gas, i.e. the throwing power. It should be pointed out at this stage that realistic calculations of the degree of ionization occurring in the cathode dark space and negative glow regions of the discharge are difficult and that several different assumptions have been applied. For example, some calculations (Francis, 1956) have assumed that the behaviour of the electrons is similar to that in a free fall situation and have used the differential ionization coefficient to calculate the ionization in the cathode fall region. Others [Drugresteyn and Penning, 1940] have used the mobility concept and Townsend's first ionization coefficient in spite of the fact that these quantities apply only to homogeneous electric field conditions. Since it is known that the field in the cathode fall region is strongly inhomogeneous, especially in the case of the abnormal glow discharge, these calculations give only very approximate results. Recent experimental observations and calculations have indicated that if all the ion current is carried by metal vapour atoms, a maximum value of less than 1% is obtained for the number of ionized particles (Spalvins, 1980). Since a significant percentage of the ion current is still carried by the inert gas ions during the ion plating process, this indicates that the number of ionized particles may be less than 0.1%. Further experimental measurements of the total current to the substrate (cathode) during the ion plating process have led to the conclusion that the number of ionized particles ranges from 0.03% to 0.1% under a wide range of conditions (Delcea, 1978). This has also led to another conclusion that in the ion plating process, the substrate is coated with a small number of energetic metal ions and a large number of energetic metal neutrals which result from collisions in the dark space (Teer, 1975). The number of ionized atoms can play a vital role in reactive ion plating, especially in low temperature reactions. Matthews and Teer (1980) reported that the minimum number of ionized atoms required to produce a hard cohesive titanium nitride (TiN) should be 0.2% to 0.3%. Furthermore, the increase in the ionization promotes very dense structures and high hardnesses (Ahmed, 1986).

2.6 THE ENERGY DISTRIBUTION OF ION AND NEUTRAL FLUXES OF GAS IN ION PLATING PROCESS

In view of the complexity of the discharge processes occurring in ion plating, it is obviously of interest to measure the energy distribution of gas ions or neutrals reaching the cathode, particularly under evaporation conditions. A knowledge of this distribution would not only allow one to gain a better understanding of the processes occurring at the cathode surface, which consequently affects the characteristics of the coating, but would also shed some light on the origin of the ions traversing the cathode dark space (Ahmed, 1980). During the last few decades, a considerable amount of information concerning the energy distribution of gas ions, in the absence of evaporant, have been published. For example, Chaudri and Oliphant (1932) used a combination of a retarding lens and a 127° electrostatic analyser to measure the energy distribution of the positive ions passing through a small aperture in the cathode of a glow discharge. They found that the energy distribution was a function of pressure, but was independent of the discharge current. They also observed that the maximum energy of the positive ions was equal to the discharge voltage.

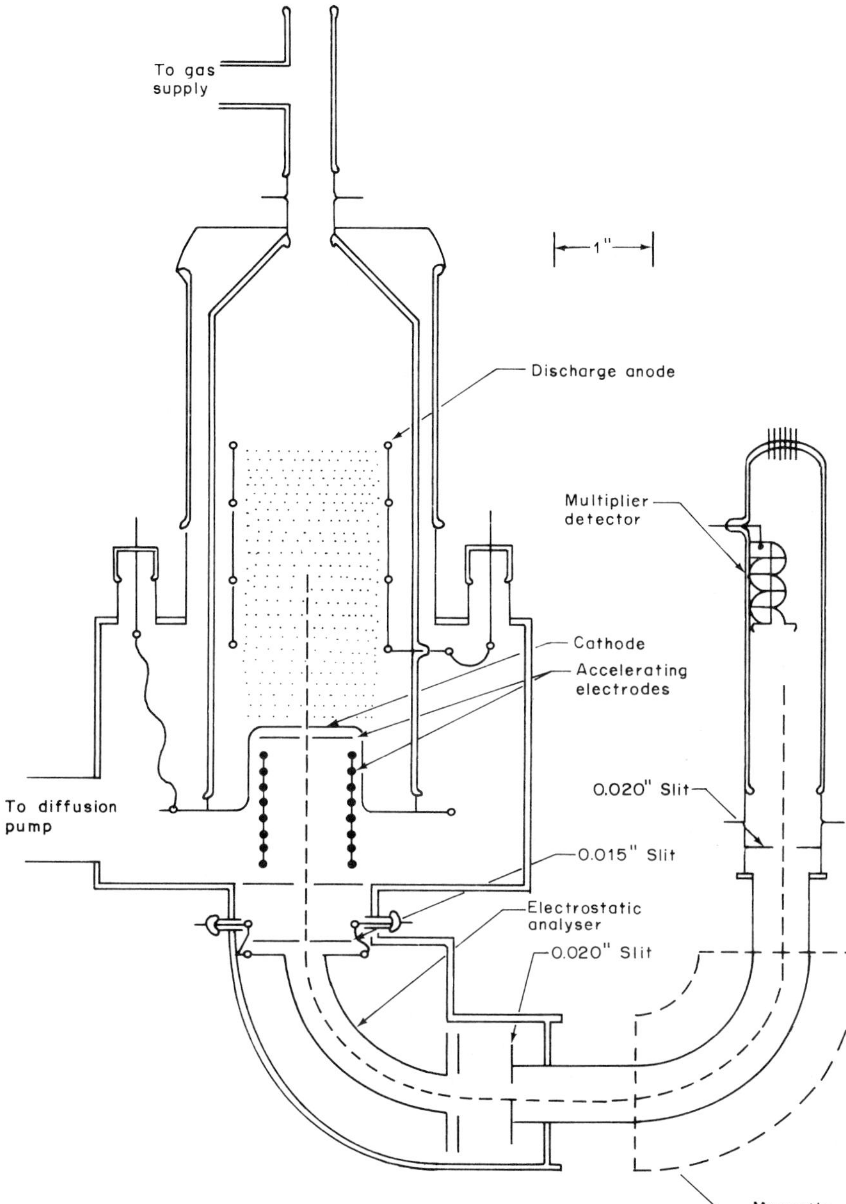

Figure 6. Experimental apparatus for the energy and mass analysis of ions bombarding the cathode in a d.c. discharge (from Davis and Vanderslice, 1963). (*Reproduced by permission of the American Physical Society*)

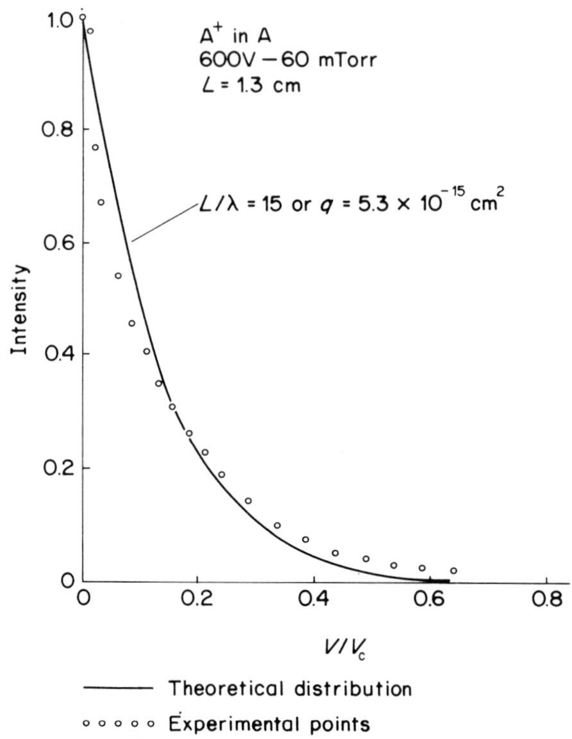

Figure 7. Energy distribution for Ar^+ ions from an argon gas discharge (from Davis and Vanderslice, 1963). (*Reproduced by permission of the American Physical Society*)

Theard (1968), Yarnold and Bolton (1949), Harrower (1955) and Hutchinson (1962) have all used an electrostatic parallel plate energy analyser to measure the energy distribution in Ar^+ ions extracted from a non-magnetic, hot cathode, gas discharge ion source. However, the most relevant work to glow discharge processes was carried out by Davis and Vanderslice (1963) who presented a theory of ion energy distribution in a glow discharge. The main conclusions of the theory have been confirmed experimentally (Houston and Uhl, 1971; Armour *et al.*, 1984). The apparatus they used is shown in Figure 6. Some of the ions striking the cathode pass through a tiny hole into a much lower pressure region where they are energy analysed and then mass analysed. The energy distribution of ions striking the cathode would be influenced not only by charge exchange but also by ionizing collisions in the sheath. On the basis of the experimental result shown in Figure 7, Davis and Vanderslice drew a model which is based on the following assumptions—see Figure 8.

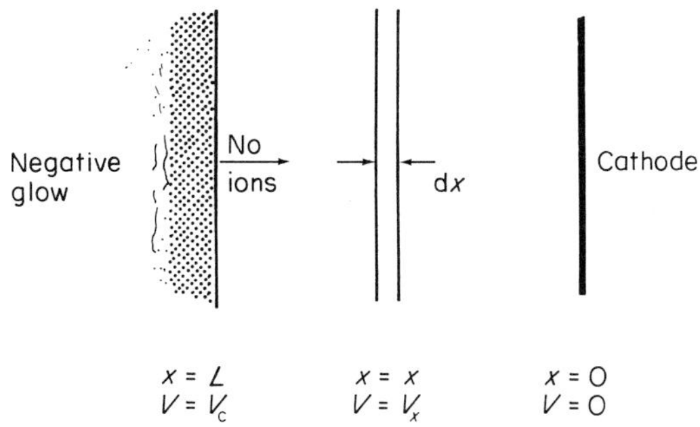

Figure 8. Model used to derive energy distribution (from Davis and Vanderslice, 1963). (*Reproduced by permission of the American Physical Society*)

(1) All ions originate in the negative glow or close to it, which means little or no ionization in the sheath.
(2) Resonant charge transfer represents the only significant energy loss mechanism.
(3) The charge exchange cross-section is independent of energy.
(4) The electric field across the sheath decreases linearly to zero at the dark space–negative glow interface.

However, the last assumption is not strictly true due to the presence of uniform positive ion space charge in the sheath. Using these four assumptions and parameters, Davis and Vanderslice obtained the following relation:

$$\frac{V_c}{N_0}\frac{dN}{dV} = \frac{L}{2\lambda}\left(1 - \frac{V}{V_c}\right)^{-1/2} \exp - \frac{L}{\lambda}\left[1 - \left(1 - \frac{V}{V_c}\right)^{1/2}\right] \tag{2.7}$$

where V_c is the target voltage, dN is the number of ions arriving with energies between eV and $(eV+edV)$, L is the dark space thickness, λ is the charge exchange mean free path, and N_0 is the number of ions leaving the negative glow. When $\lambda \ll L$ this distribution function reduces to:

$$\frac{V_c}{N_0}\frac{dN}{dV} = \frac{L}{2\lambda}\exp\left(-\frac{L}{2\lambda}\frac{V}{V_c}\right) . \tag{2.8}$$

By assuming that the theoretical cross-section is 5.3×10^{15} cm^2, theoretical results were calculated for argon and showed a reasonable agreement with experimental values as shown in Figure 7. The good agreement between the theoretical and experimental results is taken as confirmation that there is little or no ionization in the sheath. However, this theory did not cover the effect of different pressures and discharge voltages on the energy distribution which are important in the ion plating process.

Theory of Davis and Vanderslice was further confirmed experimentally by Houston and Uhl (1971) and other workers (Machet *et al.*, 1983) who found that the theory agreed with the main features of the measured energy distributions.

Based on the Davis and Vanderslice model, Teer (1977) calculated the total and average energy of ions arriving at the cathode during the ion plating process. He assumed that the mean free path and the length of the dark space are 0.5 cm and 10 cm respectively under the typical ion plating conditions, i.e. 3 kV bias and 1 Pa gas pressure. The expression achieved is:

$$\text{Total energy} \simeq \frac{2eN_0 V_c \lambda}{L} \tag{2.9}$$

$$\text{Average energy} \simeq \frac{2e V_c \lambda}{L} \tag{2.10}$$

where N_0 is the number of ions per unit time leaving the negative glow and e is the electron charge.

According to Teer, if $\lambda/L = \frac{1}{20}$, then the average energy of the ions is 10% of the maximum possible energy and the total energy carried by the ions to the cathode is also 10% of the energy dissipated in the system. The energy lost by the ions is transferred to neutral atoms. Thus energetic neutrals account for 90% of the energy dissipated in the glow discharge. This simplified model has led to the conclusion that ion plating appears to be the deposition of a small number of energetic ions and a much larger number of energetic neutrals. The most probable energies involved are an order of magnitude less than the maximum possible energy, which is equal to the discharge voltage. For example, in an argon discharge of 3 kV bias, the most probable energy of the arriving ions at the cathode is approximately 300 eV and that of the neutral particles is 135 eV.

Experimental results reported by the author (Ahmed, 1980) and other workers (Armour *et al.*, 1984) confirmed some of the above conclusions. The experimental results showed that the most probable energy of argon positive ions reaching the cathode ranges from $\frac{1}{6}$ to $\frac{1}{2}$ of the discharge voltage, depending on different parameters including the discharge pressure. Furthermore, since energetic neutrals start life as ions the energy distributions of neutrals were observed to be shifted to lower energies (Armour *et al.*, 1984). The neutral particles are mainly responsible for the bulk of the energy deposited at the cathode during ion plating. Table 1 and Figure 9 exhibit the summary of the most probable energies of argon positive ions as measured by various workers using different analytical techniques. It is of interest to mention at this point that the major effects of ion bombardment during condensation will not depend on the average particle energy but are related to the energy of individual particles (Schiller *et al.*, 1977).

Table 1.

Example	Reference	Gas	Pressure (Pa)	Ar^+ ion energy (eV)	Discharge potential (V)
1	Ahmed (1974, 1980)	Ar	1.33	900	1414
		Ar	1.2	450	940
				520	952
				600	990
				850	1128
2	Teer (1977)		Theoretical	300	3000
3	Machet *et al.* (1977, 1983)	Ar	4.9	450	3000
4	Armour *et al.* (1984)	Ar	4	430	1500
				560	2000
				700	2500
				900	3000

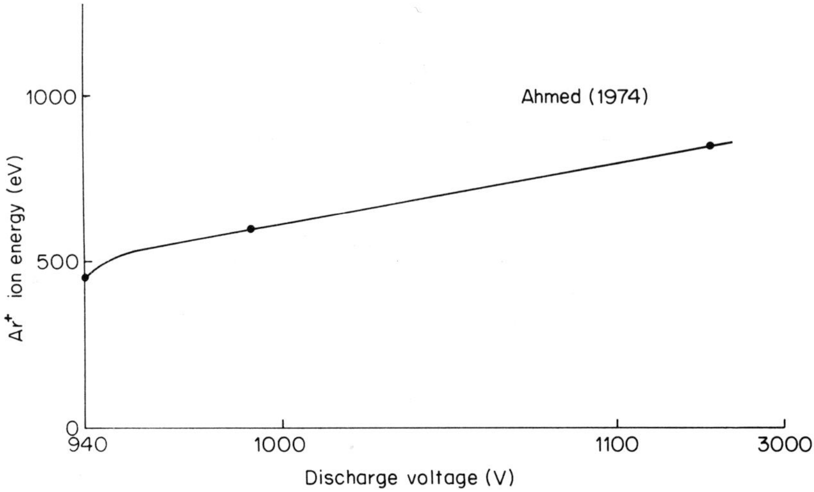

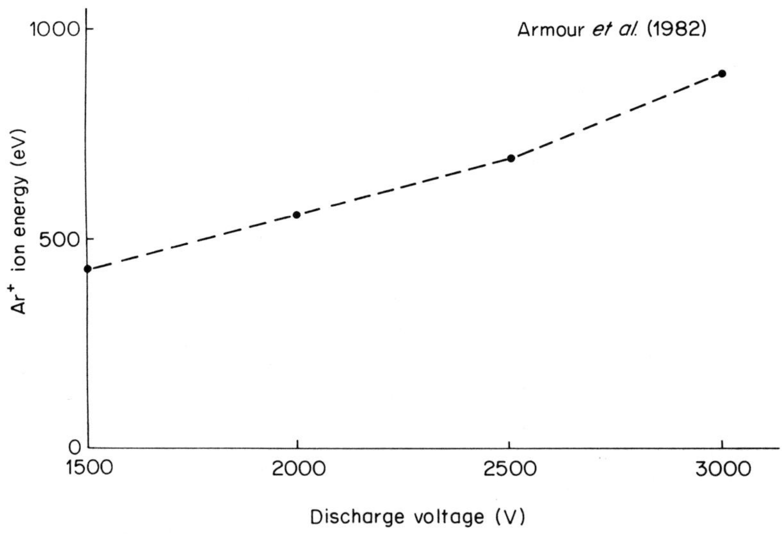

Figure 9. The energy distribution of Ar^+ ions as achieved by different workers

2.7 ENERGY DISTRIBUTION OF EVAPORATED MATERIAL

As we have seen from the previous section, there is an abundance of information concerning the measurement of energy spread of gas ions or neutrals reaching the cathode of an abnormal glow discharge. However, there has been little information concerning the energy distribution of ions reaching the cathode

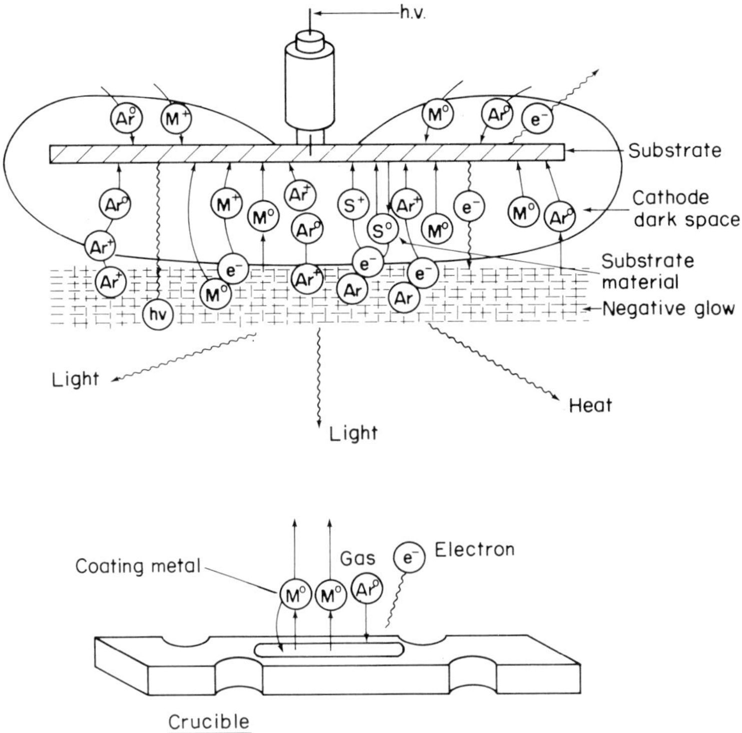

Figure 10. Schematic representation of the processes occurring in the ion plating environment

with either an abnormal or normal discharge during evaporation of a metal, i.e. during the actual ion plating process. A knowledge of this distribution would assist in a better understanding of the plasma–surface interaction processes and would assist in the control of the processes occurring in an 'ion plating' deposition system which eventually lead to good and reproducible coatings (Armour, 1982; Valizadeh, 1982). During the ion plating process, when film atoms are injected into the discharge typically from a resistively heated filament (Figure 10) a number of possibilities may occur. Film atoms may be:

(1) Scattered back to the source or to other portions of the system.
(2) Ionized by an electron–atom collision:

$$e^- + M^0 \rightarrow M^+ + 2e^-$$

where M is the metal atom, or by metastable collision

$$M^0 + G^* \rightarrow M^+ + G^0 + e^- \text{(Penning Ionization)}$$

then accelerated back to the substrate.

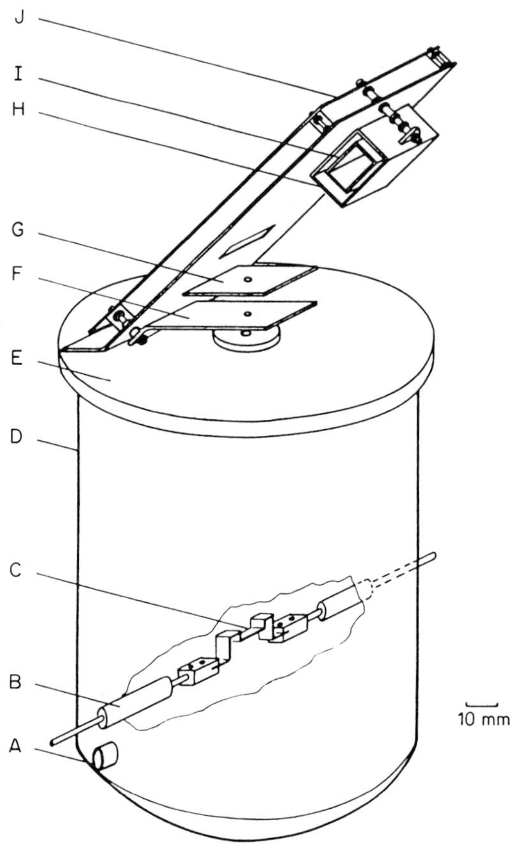

Figure 11. Experimental arrangements to study the energy distribution of argon and metal ions during the glow discharge: A, gas inlet; B, tungsten feed-through; C, boat; D, discharge tube; E, cathode; F, electron suppressor; G, earthed slit; H, screened shield; I, Faraday collector; J, parallel plate energy analyser. The discharge tube and analysing system are placed inside a main vacuum chamber (Ahmed, 1974)

(3) Struck by other film atoms in the gas phase and nucleate into fine particles (gas evaporation) which become negatively charged in the plasma and may be collected on a positively charged surface or lost to the system walls.
(4) Deposited as neutrals on the substrate (cathode).

The last possibility seems to be the most predominant in ion plating due to the charge transfer collision that was proposed by Davis and Vanderslice theory. Teer (1977) has estimated the ratio of the number of energetic neutrals to the

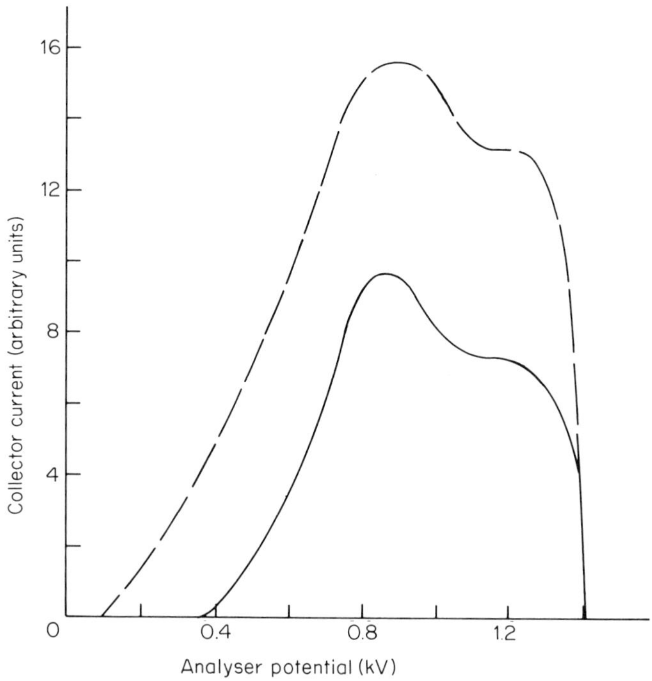

Figure 12. The energy distribution of ions (. . . . . . .) during Ar discharge at 1.33 P and 1414 V and after the evaporation of lead (———) in the Ar discharge (Ahmed, 1974)

number of energetic ions arriving to the substrate to be 14:1. The energy distribution of metal ions and neutrals during the actual ion plating process has been studied by different workers. Perhaps the first attempt to measure the energy distribution of evaporant in an ion plating process was carried out by the author (Ahmed, 1974) who used a parallel plate energy analyser to measure the energy distribution of positive ions prior to and during metal evaporation. The apparatus used is shown in Figure 11. The results obtained using this set-up are shown in Figure 12 and indicate that while the average energy of metal ions reaching the substrate is higher than that of argon ions, the most probable energy of metal ions is slightly lower. Such a decrease might be caused by the rise in pressure during evaporation of metal which leads to a depression in the energies of the evaporant ions and neutrals (Armour *et al.*, 1984). This phenomenon often occurs in conventional ion plating and in certain cases may sustain the plating discharge without the need for a carrier gas. Furthermore, it was found that the most probable energy of the evaporant ions occurred at half the applied discharge voltage, which represents the maximum energy of ions traversing the dark space if there are no collisions. Machet *et al.* (1977, 1983) used the simple experimental set-up shown in Figure 13 to determine the energy distribution of ions striking the substrate during ion plating.

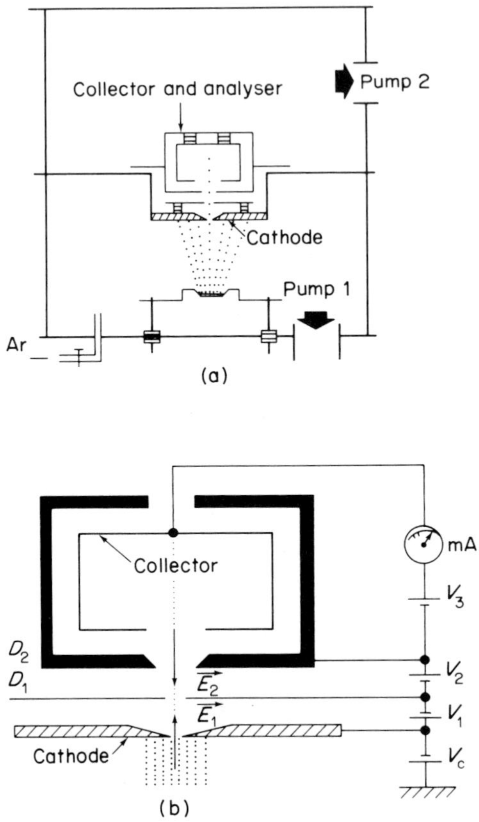

Figure 13. Experimental set-up used to measure the energy distribution of ions showing the collector and analyser assembly (from Machet *et al.*, 1983). (*Reproduced with permission from Vacuum vol. 33, ©1983 Pergamon Journals Ltd*)

The ion energy was measured by the retarding field method. Their results showed that the average energy of evaporant ions is higher than that of the argon ions by a factor of two. However, due to the simplicity of the design and the lack of a mass analyser it was difficult to interpret the results in terms of ion species and the ratio of ions to neutrals in the discharge. On the other hand, a comprehensive study has been carried out by Armour *et al.* (1984) to achieve better understanding of the processes occurring during ion plating using a highly accurate system. The system used is shown in Figure 14 and comprises an extraction/collimation arrangement followed by deflectors and electron bombardment ionizer, a retarding lens–hemispherical electrostatic analyser and a quadrupole mass analyser. The observed energy distribution shown in Figures 15 and 16 indicates that the average and the most probable energies of the evaporant ions are higher than that of the argon ions. Furthermore, the most probable energy of the evaporant ions (Cu^+) occurred at less than half the discharge voltage, while the most probable energy of neutrals (Cu^0) occurred at less than one-third of the discharge voltage. This result reflects the importance of charge exchange collisions in the dark space. Consequently, when assessing

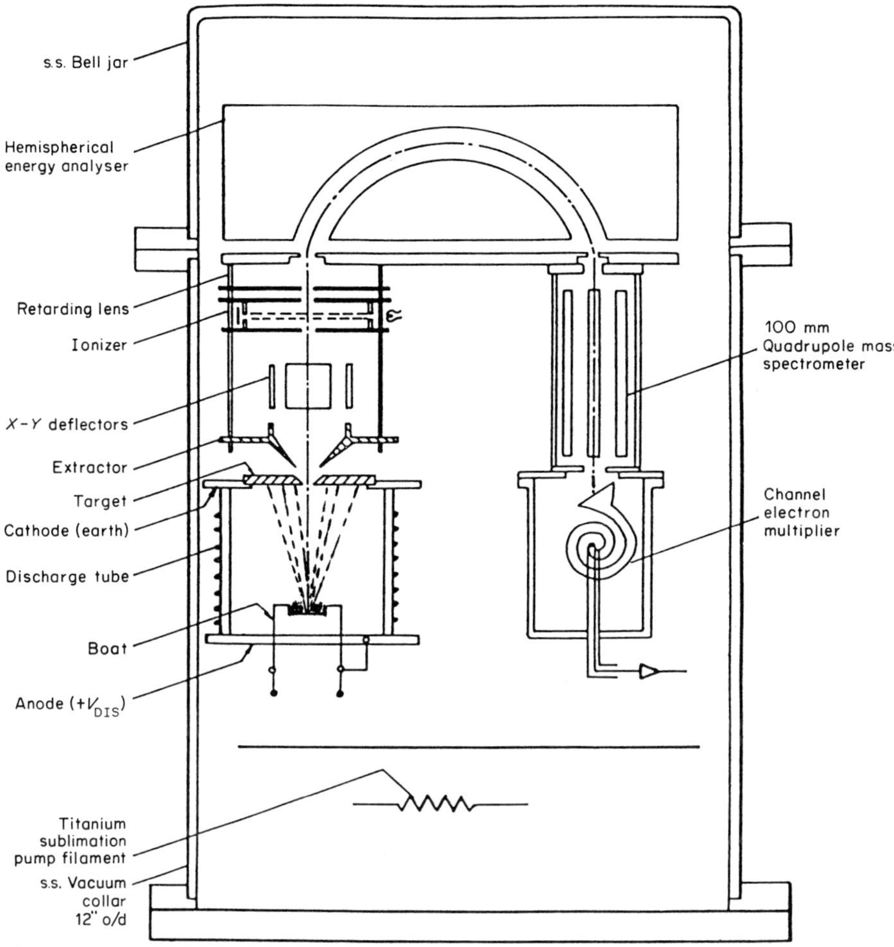

Figure 14. Experimental ultra-high vacuum ion plating and analysis system (rom Armour *et al.*, 1984). (*Reproduced with permission from Vacuum vol. 34, ©1984 Pergamon Journals Ltd*)

the relative importance of the different processes associated with energetic particle bombardment during ion plating, it is necessary to consider energies which are typically below 1 keV (Armour *et al.*, 1984). Furthermore, the high ratio of neutrals to ions and their associated high energies implies that neutral particles are responsible for the bulk of the energy deposited at the substrate during ion plating.

In order to display the results of energy distribution of evaporant ions in an ion plating process in a general form, Table 2 and Figure 17 shows the results achieved by different authors using various analysing techniques. It should be emphasized at this point that the measurements of the energy distribution depend

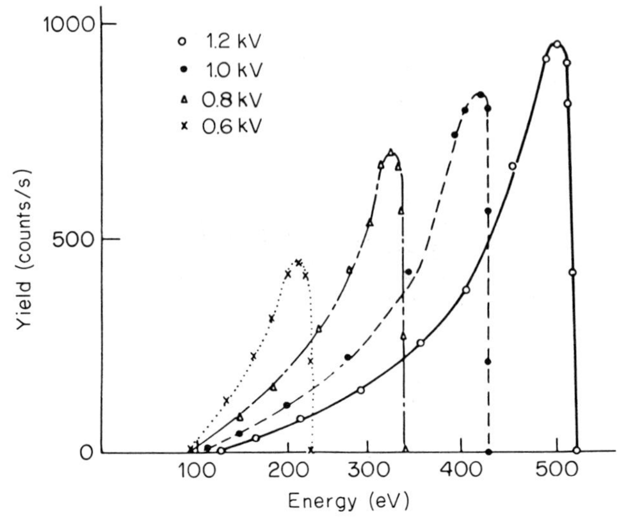

Figure 15. Energy distribution of Cu^+ ions leaving an argon supported discharge for various discharge voltages (from Armour *et al.*, 1984). (*Reproduced with permission from Vacuum vol. 34, ©1984 Pergamon Journals Ltd*)

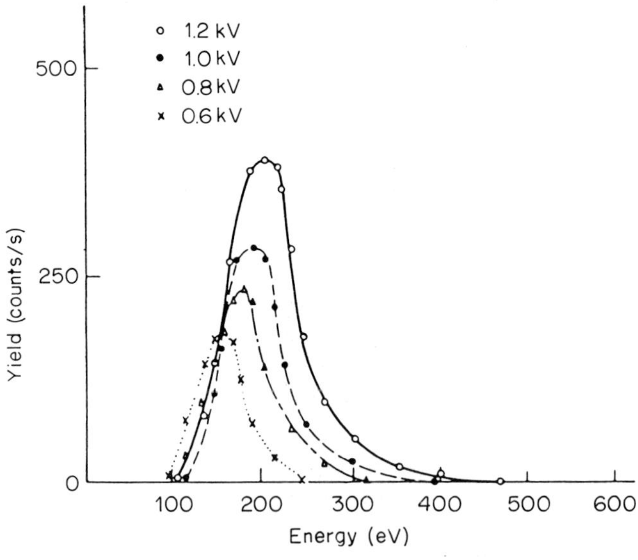

Figure 16. Energy distribution of Cu^0 neutral particles leaving an argon supported discharge for various discharge voltages (from Armour *et al.*, 1984). (*Reproduced with permission from Vacuum vol. 34, ©1984 Pergamon Journals Ltd*)

Table 2.

Example	Reference	Gas	Pressure (Pa)	Film Material	Metal ion energy (eV)	Discharge potential (V)
1	Ahmed (1974, 1982)	Ar	1.33	Silver	850	1414
2	Teer (1977)		Theoretical	—	300	3000
3	Machet *et al.* (1977, 1983)	Ar	4.9	Silver	1100	3000
4	Armour *et al.* (1984)	Ar	>4	Copper	200	600
					320	800
					420	1000
					500	1200

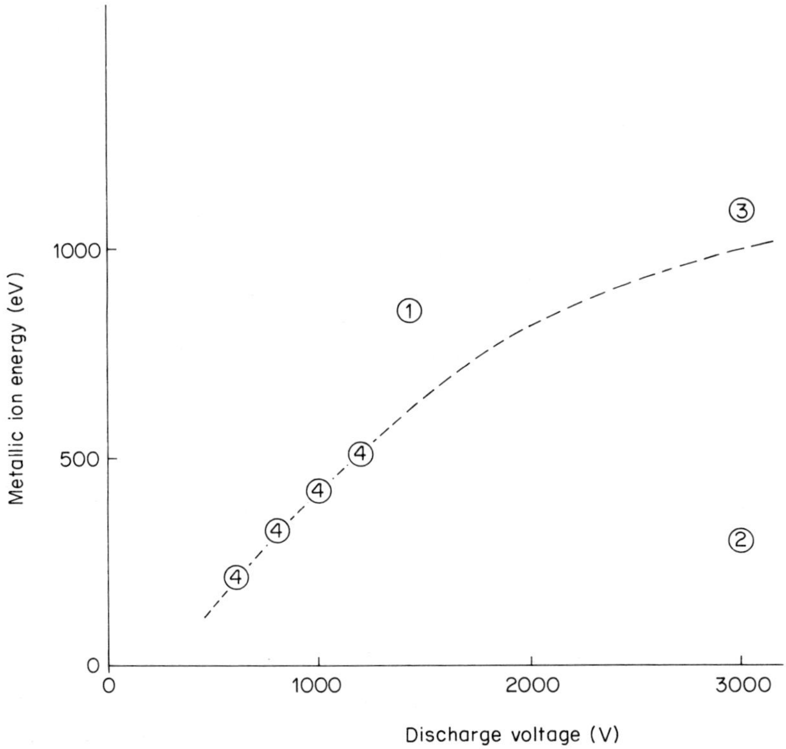

Figure 17. Energy distribution of metal ions as achieved by different workers. (1) Ahmed (1974); (2) Teer (1975); (3) Machet (1983); (4) Armour (1984)

on many factors including pressure and the geometry of the analyser, and therefore the results shown in Table 2 and Figure 17 are only meant for demonstration. However, it does not seem to be unreasonable to suggest that the average energy of arriving evaporant ions and neutrals at the cathode ranges between ⅓ and ½ of the discharge voltage.

2.8 PROCESSES OCCURRING AT THE SUBSTRATE SURFACE

From the previous sections it is apparent that, during the film deposition period, the substrate is bombarded with high energy ions and neutrals of the carrier gas and film material, particles having thermal energy as in normal vacuum evaporation as well as some Auger electrons. Consequently, the early stages of film growth are controlled by a complex mixture of processes involving collision, penetration and entrapment of high energy particles (Walls *et al.*, 1978), sputtering of loose particles condensing on the surface, reburial of surface film particles by previously sputtered substrate ions, and enhanced diffusion due to significant radiation damage produced by the continuous ion bombardment.

Knock-on sputtering effects (recoil implantation) in which surface atoms are knocked into, rather than out of, the substrate may also be important even though the penetration depth is very small (Armour *et al.*, 1980; Carter and Armour, 1981; Hurley, 1981; Nelson, 1977; Holland, 1975; Mattox, 1979). In addition, the glow discharge acts as a rather intense source of ultraviolet radiation which can affect organic surfaces and possibly insulator surfaces (Mattox, 1982b). However, it was suggested by Armour *et al.* (1980) that since the energies during ion plating are typically below 1 keV, this implies penetration ranges of only tens of Ångströms. When this factor is combined with the fact that in the ion plating process, the film growth rate is higher than the sputtering rate caused by ion bombardment, it becomes apparent that direct ion implantation of film species and the deposition of energetic film particles play only a comparatively minor role in the actual material growth. Thus the deposition of energy (which is the transfer of energy from energetic neutrals and ions to the film and substrate atoms) and the associated mixing process represents the major contribution of the energetic particles (Carter and Armour, 1981).

In view of this, therefore, the most important ion–solid interactions which affect coating characteristics such as adhesion and hardness will be discussed in detail in the following sections.

2.8(a) Interface Formation

The adhesion between a film and its substrate depends to a large extent on the condition of the interfacial layer that is formed between them (Mattox, 1965). Interface regions may be classed as abrupt, compound, diffusion, pseudo-diffusion, irregular or a combination of these (Pulker, 1984). Figure 18 shows

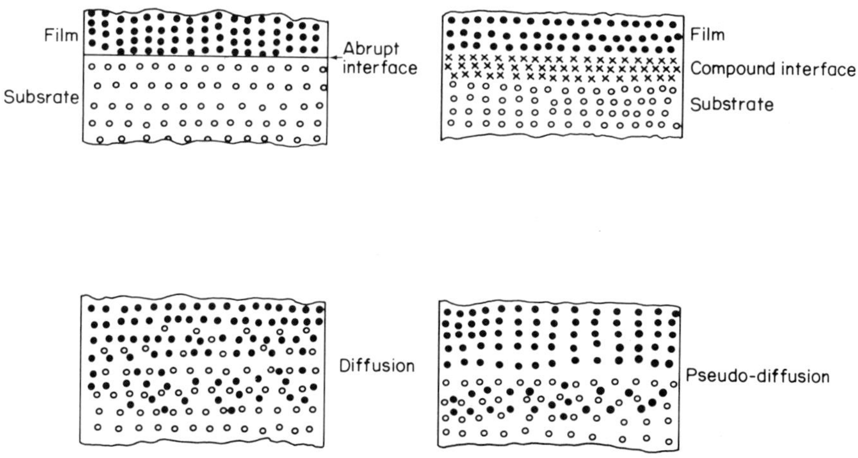

Figure 18. Schematic representation of the types of interface regions between film and substrate

the different types of interfaces. Abrupt interfaces form when there is little chemical reaction between the coating and substrate materials. Compound interface regions form when the materials react to form intermetallic or chemical compounds, and diffusion interfaces form when they have considerable solubility with one another. This interfacial layer is characterized by a gradual change of the lattice and the composition within the film–substrate transition areas. Pseudo-diffusion interfaces can be formed by implantation processes at high particle energies or in the case of sputtering and ion plating by gas-phase backward scattering of the sputtered substrate material atoms, thus mixing coating material with the substrate (Mattox, 1965). Pseudo-diffusion can form graded interfaces between materials that do not mutually diffuse. Irregular interfaces can be formed by bead blasting or etching the substrate.

The significant adhesion between the coating and the substrate in ion plating is mainly related to the formation of graded diffusion or pseudo-diffusion interfacial regions of slowly changing composition. However, it has been argued that such interfaces may arise either from direct penetration and implantation of the energetic coating atoms into the substrate then into the growing coating or from recoil implantation of coating atoms into both regions via collisions with energetic incident flux (Carter and Armour, 1981). Remembering that the incident particles will have energies below 1 keV, it is expected that the penetration depths will be a few atomic layers. Such penetration is not capable of transferring substrate atoms into the coating and thus providing broad interfaces with gradual density gradient of coating and substrate atoms. Therefore, direct or recoil implantation cannot explain the formation of broad interfaces which may reach sometimes a few microns (Armour *et al.*, 1980) and there is good evidence that atom movement by diffusion is an important process to explain such effects. In ion plating, thermal interdiffusion between the coating and substrate atoms may occur if the substrate temperature is increased as a result of the power input from the discharge and the heat radiation from the evaporation source. The diffusion rates may also be enhanced by the creation of an increased population of lattice defects resulting from the recoil events between projectiles and target atoms and between target atoms which are displaced by projectile target atoms thus creating highly mobile defects at high substrate temperature. Such an enhancement, which is called radiation enhanced diffusion, can promote deep penetration of the coating into the substrate thus forming a graded interface. Nelson (1977) suggested that even with a number of atomic displacements of 10^{-2} atom $^{-1}$ s^{-1}, radiation enhanced diffusion will be several orders of magnitude higher than thermal diffusion at substrate temperatures of the order of a few hundred degrees Celsius, typical of most deposition processes in ion plating. Furthermore, Eltoukhy and Green (1980) have produced experimental evidence that the measured diffusion coefficients in ion plating are five orders of magnitude greater than the thermal values of InSb/GaSb multi-layer structure due to radiation enhanced diffusion. It has also been suggested by Hurley (1981) that radiation enhanced diffusion is most effective at temperatures where vacancies become mobile, although interstitials may diffuse at lower temperatures. However, the enhanced diffusion process will be limited by defect recombination at lower temperatures and by losses to dislocation or other sinks at higher temperatures.

In the absence of enhanced thermal diffusion, the generation of target (substrate or coating) atom recoils due to mutual collisions between target atoms and between projectiles and target atoms, albeit with low energy, can move such atoms over short distances. If the frequency of such events is high then initially the atoms of the substrate and the film and later the film atoms alone may mix rather effectively and lead to some interface broadening and coating homogenization (Carter and Armour, 1981). This process is known as cascade mixing to distinguish it from recoil mixing which involves collisions between projectile and target atoms only. Recent studies have shown that even at low energy projectiles the cascade mixing processes can be more effective in atom redistribution than recoil mixing alone (Littmark and Hofer, 1980). Based on this, Armour *et al.* (1980) suggested a model to calculate interface broadening in ion plating. According to their model, effective diffusion coefficients of the film and substrate atoms during film growth are considered to be responsible for such broadening. The depositing film is assumed to act as a plane source from which atoms diffuse into the substrate with a diffusion coefficient (Dx) which depends on the energy deposition at the interface (see Figure 19). By applying the diffusion approximation to the ion plating process, Armour *et al.* (1980) obtained the following expression for the total interfacial broadening ΔX_T (inward and outward broadening).

$$\Delta X_T \simeq \left(\frac{J \ell^2 \overline{\theta} (E, \theta)}{E_d \, C \, J_T} \right)^{1/2} \qquad (2.11)$$

where J is the flux density of energetic ions and neutrals of both the depositing material and the carrier gas, ℓ is the mean value of distance travelled by recoil before coming to rest, λ is some small characteristic distance, $\overline{\theta}$ (E, θ) is the energy deposited per unit depth at the interface by atoms arriving at the surface with energy E, E_d is the effective displacement energy, C is the

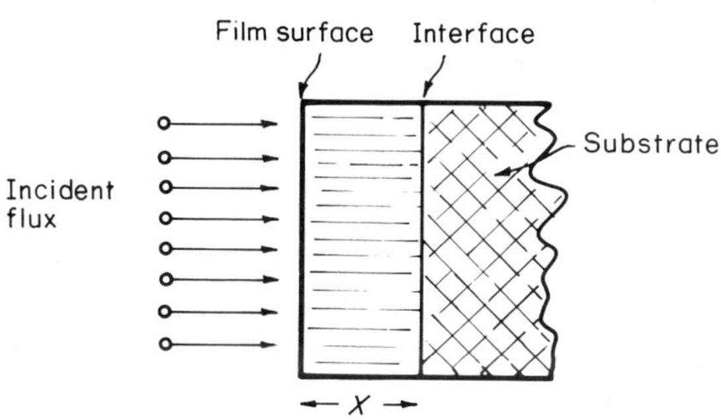

Figure 19. Model used to calculate interface broadening in ion plating. (D G Armour, Private Communication)

effective sticking probability and J_T is the flux of more or less thermal energy particles.

Therefore, for a typical plating employing an argon discharge operating at a voltage of 2 kV and a current density of 1 mA cm^{-2} of substrate surface area, a net film growth rate of 600 Å min^{-1}, i.e. 10 Å or approximately 2×10^{15} atoms cm^{-2} s^{-1} implies a value of approximately 2 for a ratio j_T/J. Since for these conditions, the majority of the energetic particles have energies of about 500 eV then one would expect the following values for the other parameters, $\ell \simeq 2$–3 atomic spacings, $\lambda \simeq 2$–3 atomic spacings and $\overline{\theta} \simeq 100$ eV/atom layer, E_d in metals is approximately 25 eV (Carter and Grant, 1976). Hence

$$\check{X}_T = \sqrt{30} \qquad \simeq 6\varrho \tag{2.12}$$

where ϱ is the interatomic spacing.

The cascade mixing model thus predicts that the film–substrate interface may typically be broadened by approximately 6 lattice spacings or about 20–30 Å. Interface regions of an order of 40 Å were measured experimentally by the author for silver coating on copper substrate (Ahmed, 1974). The model also predicts that the broadening should be maximized by using high discharge voltages and low deposition rate. It should be emphasized that this model is oversimplified since the effective diffusion coefficient will vary with depth into the substrate and within the film, as well as with time.

In a further attempt, Carter *et al.* (1981) extended the diffusion approximation to derive a criterion that significant atomic redistribution can occur during coating growth in ion plating. They suggested that 'physical mixing' is of more general significance during ion plating than thermal or radiation enhanced diffusion which is specific to the substrate–coating material combination. By assuming that each projectile creates a number of recoiling displaced atoms which have low kinetic energy and which will travel only a short distance before coming to rest, Carter *et al.* (1981) showed that the ratio of atomic mixing velocity $\nu_M(E)$ to sputtering boundary velocity $\nu_S(E)$ is

$$\frac{\nu_M(E)}{\nu_S(E)} = \frac{1}{6} \frac{\nu(E)\,\lambda\,(E)}{Y(E)\,R\,(E)} \tag{2.13}$$

where $\nu(E)$ is the number of recoiling displaced atoms created by a projectile at energy (E) in a target over a depth $R(E)$, $\lambda(E)$ is the mean value of distance travelled by recoils before coming to rest, and $Y(E)$ is the sputtering field. For low energy projectiles, $\lambda(E)$ is expected to be a function of $R(E)$ (e.g. $0.1 - 0.5$. R(E)). Estimates of $\nu(E)$ and $Y(E)$ may be made from radiation damage and sputtering theory and were evaluated by Andersen (1979) to be in the region of 0.1 (for coinage metals) to 0.6 (for elemental semiconductors). Therefore, $\nu_M(E)/\nu_S(E)$ is expected to be in the range of $(0.01 - 0.2)$. This indicates that the rate at which substrate and coating atoms and later coating atoms alone intermix is generally substantially lower than the sputter erosion rate. However, this assumption underestimates the velocity ratios since at projectile energies greater than about 1 keV the ratio of $\nu(E)$ to $Y(E)$ will increase slowly with energy. Furthermore, experimental studies of intermixing during sputtering (Andersen, 1979; Etzkorn *et al.*, 1980) and interface broadening suggest (Tsaur *et al.*, 1979) that intermixing rates may exceed those derived from the diffusion approximation by an order of magnitude. Therefore, Carter *et al.* (1981) concluded that conditions can be obtained where boundary mixing velocities can occur at rather similar rates to those of sputter erosion of the surface. This has an important consequence since if, relatively strong atomic redistribution (physical mixing) with substrate coating interpenetration and good adhesion are to be achieved then the net rate of coating should be relatively slow, representing a small balance between deposition and sputtering (and consequently mixing) processes. It must be emphasized that the above criterion concerns the general three-dimensional irradiation enhanced diffusion of atoms in the solid. It is known (Luzin and Tyusher, 1979), however, that surface atomic motion may be even more enhanced by irradiation since the collisional energy required to stimulate surface migration across a low energy barrier may be lower than that required to create displacement recoils and bulk migration.

2.8(b) Gas Entrapment

The continuous bombardment of the substrate surface with inert gas prior to and during film deposition normally results in the entrapment of gas atoms in the coating and substrate. The entrapment process occurs if the incident particle kinetic energy is sufficient to allow penetration beyond the coating or substrate surface plane (Carter and Armour, 1981). This effect has long been recognized and studied and forms the basis for inert gas pumping by 'getter ion pump' type vacuum pumps. Furthermore, inert gases have been deliberately incorporated into surfaces and films by low energy ion bombardment for a variety of reasons including surface thermometry (^{85}Kr), diffusion studies (H, He) and long term storage of radioactive inert gas isotopes (Mattox, 1982b). Mattox and Kominiak (1971) have reported the incorporation of helium into gold using low energy ion bombardment to concentrations as high as 40 at% in spite of the fact that the measured solubility of helium in gold is zero. In general, it has been found that the ability to incorporate the insoluble gases into the lattice of the surface or depositing film depends on many factors including the mobility, trapping sites, and temperature as well as the energy of the depositing species. For light inert gases the entrapment occurs at lattice vacancies, whilst for heavier ions non-substitutional dissolution may also occur. Thus this process is particularly effective in the case of amorphous materials rather than crystalline materials (Mattox, 1982a; Carter and Grant, 1976). Simple calculations based on measured trapping coefficients in the low keV region (Winters and Key, 1967) have indicated that argon concentrations of the order of several per cent may be reached at the end of the plating run. Although this may be a high estimate due to the fact that the measurements on which it was based were carried out under radically different conditions of flux and substrate temperature, it is not unreasonable to assume that there is a considerable amount of gas trapped in the film. For example, in a typical ion plating process at 1 Pa argon pressure, the rate of arrival of gas atoms and impurities to the substrate is about 4×10^{18} particles $cm^{-2} s^{-1}$. For a typical evaporation rate of 1500–300 Å min^{-1}, the rate of arrival of gas atoms and impurities is about $(0.5-1) \times 10^{16}$ particles $cm^{-2} s^{-1}$. This is equivalent to the rate of arrival of impurity gas to the substrate at a pressure of 10^{-3} Pa. Thus, the arrival rate of impurity gas to the substrate is approximately 10^2 times greater than that of the atoms of the growing film on the substrate. It is worth pointing out at this stage that the incorporation of Ar to these fluences at not too high a temperature below vacancy migration in metals, leads to bubble formation. The presence of gas bubbles in the growing films may have significant influence on the grain size and other characteristics of the film (Armour, 1986). By increasing the evaporation rate the arrival rate of impurity gas to the substrate can be made equivalent to that of evaporated material. This is before the effect of substrate bias on the gas entrapment is taken into account. In fact Winters and Kay (1967) have measured the argon content of sputtered deposited nickel film as a function of bias and their results are shown in Figure 20, with the 0–200 V range classified in the inset. The decrease in argon content observed is because incoming argon ions at these energies have a high probability of sputtering previously sorbed argon but only a small probability of sticking. However, as the bias voltage increases to over 200 V, the energy of the incoming ions increases leading to more entrapment in the growing film. As soon as this effect dominates the sputtering effect, the argon content starts to rise again.

Gas entrapment will continue to increase with particle energy until at high energies sputtering and heating caused by ion bombardment causes the release of the entrapped gas. In an extension of their earlier work on argon incorporation in thin films Winters and Kay (1972) studied the composition of films deposited in a gas mixture of argon and nitrogen. As a result of this study, they divided metals into three classes.

(a) Metals which chemisorb molecular nitrogen and form a nitride, e.g. W.
(b) Metals which do not chemisorb molecular nitrogen but still form a nitride, e.g. Ni.
(c) Metals which neither chemisorb nitrogen nor form a nitride, e.g. Au.

Winters and Kay found that bias sputtering caused the nitrogen content of gold to increase, due to the increased sorption of nitrogen when the film was bombarded with energetic N_2^+ ions. However, although this same effect is present with the tungsten and nickel films, it is overshadowed by resputtering of the previously sorbed nitrogen, as it was with the tantalum film in pure argon at low bombardment energies. This is a consequence of the large sputtering yield of nitrogen on tungsten. Winters and Kay concluded that there is adequate

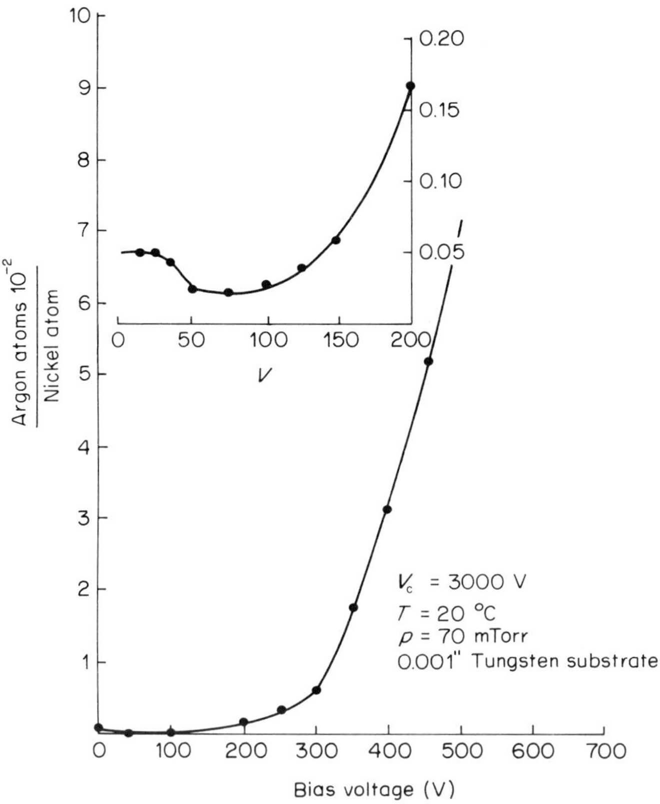

Figure 20. Argon concentration of sputtered nickel film versus bias voltage (from Winters and Kay, 1967). (*Reproduced by permission of the American Institute of Physics*)

evidence that most chemisorb gases have large sputtering yields and low thresholds. Therefore, it appears that biasing the substrate can lead to either an increase or decrease in the concentration of trapped gases depending on their surface adsorption characteristics which in turn control their sputtering yields.

Although in many applications, the incorporation of gas particles is undesirable since it may cause blistering or adhesion failure on subsequent heating, in other applications such incorporation is desirable to produce compound thin films of oxides, nitrides, carbides, and related materials (Harper, 1984).

It may be concluded therefore that inert gases can be incorporated into surfaces even though the sputtering rate is high. A similar, probably higher entrapment rate might also be expected for chemically active gases such as nitrogen and oxygen. The degree of gas incorporation depends on many factors including the coating structure which is itself a function of the bombardment rate. Any increase in substrate temperature during sputter cleaning or coating leads to an increase in the desorption rate but may also promote diffusion away from the surface (Hurley, 1981).

2.8(c) Intrinsic Stress

The durability of thin films depends mainly on their stress behaviour. In fact, almost all thin films regardless of their production methods are in a state of stress. Such mechanical stress can consist of external stress caused by external force attack, for example, in the case of poor assembly of a component supporting the film, or thermal stresses due to the difference in the thermal expansion coefficients between the coating and substrate materials. The third and the most important stress is the intrinsic stress which arises due to the accumulating effect of atomic forces generated throughout the coating volume by atoms which are out of position with respect to the minima in the interatomic force fields (Mattox, 1982b). Thermal stresses usually become important if the melting point of the coating material is lower than the substrate temperature. In this case, stress relief can occur as the coatings are brought to room temperature or annealed following deposition (Pulker, 1984). The resultant material flow can lead to the formation of hillocks or holes depending on whether the temperature change places the coating into compression or tension. This has been observed in Al, Au and Pb films prepared at elevated temperatures (Chapman, 1980).

If higher melting point materials are deposited on a substrate at lower temperatures then the intrinsic stress suppresses the thermal stress. In general, evaporated films have a tensile stress where the film tends to contract parallel to its surface (concave up), while sputter deposited films have a compressive stress where the film expands parallel to the surface (convex). Ion bombardment and gas entrapment during film growth have a strong effect on the intrinsic stress. It has been reported that for evaporated films prepared under continuous ion bombardment the stress changes from tensile to compressive (Cuomo, 1982; Hoffman and Gaerttner, 1980), while in r.f. bias sputtering the stress changes from highly tensile to compressive as the substrate bias increases (Hoffman and Gaerttner, 1980) (Figure 21). In magnetron sputtered films, Thornton and Hoffman (1981) reported the dependence of stress on gas pressure, angle of incidence and discharge current. A transition pressure (P_t) was observed below which large compressive stresses exist in metal films such as Al, Cr, Gd, Fe, Mo, Ni, Rh, Si, Ta, V, W, and Zr . Such coatings were reported to exhibit high optical reflectance and low resistivities and often contained considerable amounts of entrapped working gas. Above the transition pressure the coatings were in tension, exhibited lower reflectances and higher resistivities and contained less entrapped gas. Furthermore, it has also been demonstrated by Thornton and Hoffman that stress dependence on the angle of incidence showed a similar relationship to pressure while the increase in discharge current led to an increase in the transition pressure (P_t). This was related mainly to gas scattering of the sputtered metal species where at elevated pressures gas scattering causes the

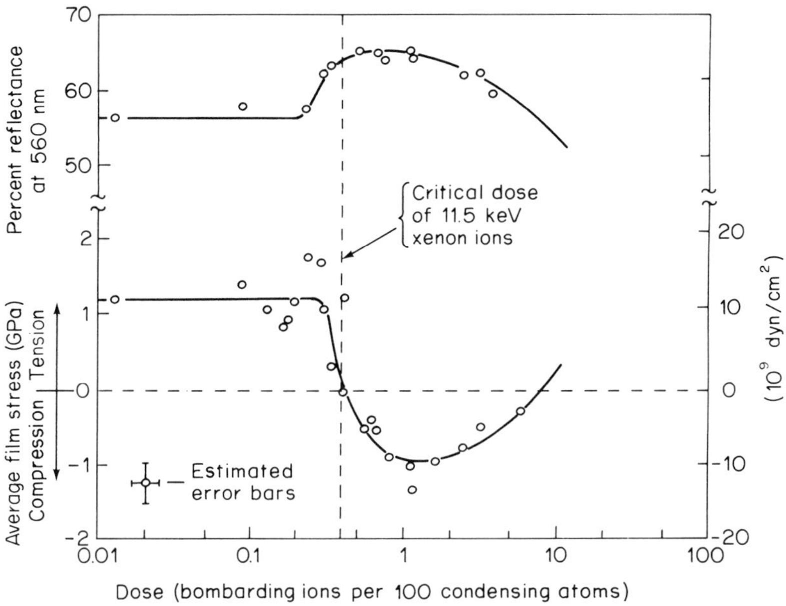

Figure 21. Average film stress and optical reflectance at 560 nm of thermally evaporated chromium films subjected to concurrent 11.5 eV xenon ion bombardment during deposition (from Hoffman and Gaerttner, 1980). *(Reproduced by permission of the American Institute of Physics)*

sputtered atoms to reach the substrate in oblique directions which promote an open structure coating of the type that yields tensile stresses. Such a structure cannot sustain a mechanical force (Thornton and Hoffman, 1981). Gas scattering can also cause the reduction of the average particle energy which is delivered to the substrate and thus affect the stress. Interpretations of intrinsic stress modifications by ion bombardment have been focused on the transfer of momentum (Hoffman and Gaerttner, 1980) or energy (Hirsch and Varga, 1980) from the ion to the film atoms. Hoffman and Gaerttner (1980) have studied the changes in the stress of evaporated Cr film as it was bombarded with 3.4 keV ar^+ ions or 11.5 keV Xe^+ ions. They reported a sharp transition from tensile to compressive stress at a critical value of the arrival rate ratio of Ar^+ or Xe^+ ions to Cr atoms. For example, at low ion energy, the ion/atom arrival rate ratio for stress relief is higher than that at high ion energy. In another study, Cuomo *et al.* (1982) reported that the highly tensile stress of evaporated Nb films was reduced to near zero after bombardment with a low flux of 100 eV Ar^+ ions. In their study they showed that gas entrapment is also influenced by ion bombardment due to preferential resputtering of gas species.

In practice, however, the compressive state of stress is often desired, as many coating materials are stronger in compression than in tension because fractures will tend not to propagate. This is important when the substrate has a greater thermal expansion coefficient than the coating (Pulker, 1984).

2.8(d) Compositional Effects

During the ion plating process, ion bombardment of the deposited film may result in compositional effects due to sputtering, ion or recoil implantation, diffusion or surface segregation (Kelly, 1980). Without diffusion, atoms will be removed from the surface either by sputtering in a stoichiometric ratio after an initial conditioning period, or by preferential sputtering which may be significant in some cases. The case of preferential sputtering is not fully understood but it is considered to be associated with segregation at the surface in some cases. Diffusion in the surface region can have a pronounced effect on composition since the presence of a high defect concentration and a high temperature will enhance diffusion such as that found with radiation enhanced diffusion. This has been discussed in detail in Section 2.8(a). However, since the surface acts as a sink for point defects, the defect flow will allow solute segregation with small ions segregating to the surface (Mattox, 1982b). If the growing film is bombarded with reactive gas species then compound thin films of oxides, nitrides, carbides could be produced (Vossen and Kern, 1978). It has also been reported that ion bombardment during deposition is important in obtaining stoichiometric composition of compounds, but excessive bombardment may lead to preferential loss of a constituent material (Ahmed and Teer, 1982; Mattox, 1982a).

2.8(e) Temperature Rise

Most of the energy of bombarding ions and neutrals is transferred to the substrate surface and appears as surface heating. Although the substrate is usually attached to a water-cooled holder in the ion plating process, with current densities as high as 0.5 mA cm^{-2}, the temperature rise may reach a few hundred degrees. Figure 22 shows the typical temperature rise in the substrate during the sputter cleaning cycle of the ion plating process. In general, the surface to mass ratio, thermal properties of the system and the energy input of the system determine the resulting bulk temperature of the bombarded substrate. In most cases the increase in substrate temperature up to about 300°C is advantageous in improving coating adhesion and structure properties in the ion plating process. For example, if diffusion is possible between film and substrate material then the rise in substrate temperature enhances thermal diffusion and therefore improves adhesion and other mechanical properties. In other cases, the increase in substrate temperature may result in the formation of a graded interface with a continuous change in chemical composition which may counterbalance the effect on varying expansion coefficients thus reducing undesired mechanical stresses. In reactive ion plating, the increase in substrate temperature may increase the chemical reaction ability hence developing dense crystalline films

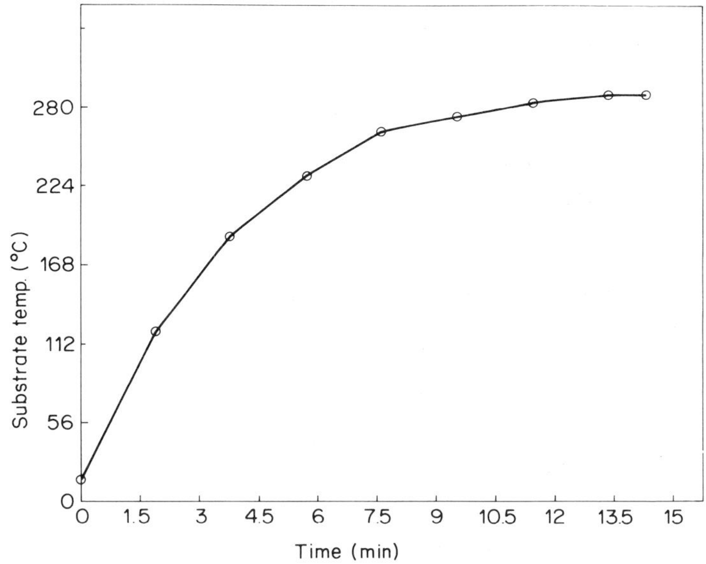

Figure 22. Temperature rise of a steel substrate during sputter cleaning at 1 Pa argon pressure, 3 kV bias and 0.3 mA cm^{-2} current density

(Pulker, 1984).However, excessive rise in substrate temperature may lead to coatings with bad adhesion and structure for certain coating–substrate combinations (Mattox, 1982b).

2.8(f) Nucleation Modes

The condensation of a vapour atom is determined by its interaction with the substrate surface. The atom is attracted to the surface by the instantaneous dipole and quadrupole moments of the surface atoms and it loses its energy in a short time if the kinetic energy is not too high (Chopra, 1969). The incident atom transfers its energy to the substrate lattice and becomes a loosely bonded 'adatom'. The adatoms then diffuse over the surface exchanging energy with the lattice and other adsorbed species, until they either desorb by evaporation or back-sputtering from the substrate surface or become trapped at low energy sites. This process eventually leads to island formation. With a continuous deposition these islands grow together to form a continuous coating (see Figure 23). This is the case in vacuum evaporation. If the condensating atoms and condensation surfaces are bombarded with energetic plasma species then ions and neutrals will transfer this energy, momentum and/or charge to the substrate or the depositing film surface. Consequently, ion bombardment will: (1) enhance the surface mobility of adatoms; (2) stimulate or accelerate the nucleation, growth of the nuclei and coalescence at the initial stage of film formation; (3) create or activate sites that stimulate nucleation processes; (4) develop nucleus

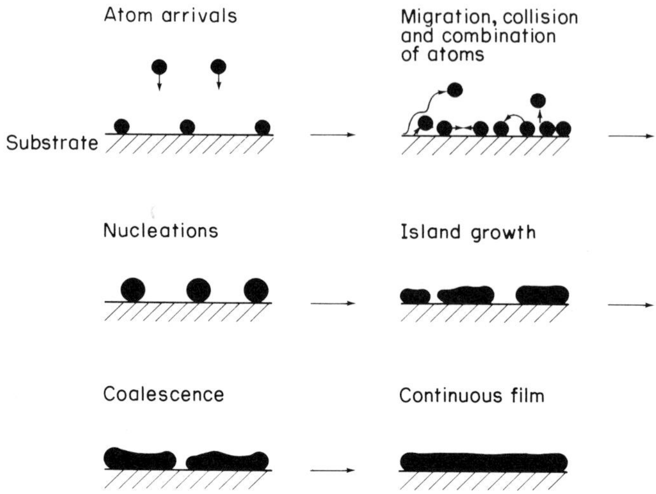

Figure 23. Schematic representation of the stages of metal film growth in vacuum evaporation

orientation on the single crystal substrate by increasing concentration of small condensation centres; (5) create a recrystallization effect of the deposited film by incident energy of ions; (6) increase the bonding energy between deposited film by incident energy of ions; and (7) decrease stress in the film. In addition, ion bombardment of the surface will tend to remove and disrupt contamination and barrier layers which prevent the diffusion/reaction nucleation model (Ahmed and Teer, 1982).

If the condensing atom reacts strongly with the surface, the surface mobility will be limited and the nucleation density will increase. This will lead to the formation of a continuous monolayer on the substrate surface (Mattox, 1982b; Nicolet, 1978). If there is a chemical reaction or diffusion, the condensing atoms will react with the surface to form compound or alloy layers which extend both normally and laterally to the surface. Such a chemical reaction may also produce strong heat of formation of a compound, which can assist diffusion and ion beam mixing.

This mode of growth will tend to decrease interfacial porosity and 'grade' the properties of the interfacial regions (Mattox, 1969).

Extensive interfacial reaction may degrade the interface by the formation of porosity due to differential diffusion rates or by high intrinsic stresses generated by thick interfacial compound layers (Vossen *et al.*, 1977). Very thin contamination layers on the surface can convert the diffusion/reaction nucleation mode to the non-reaction mode with an attendant undesirable interfacial mode.

Chapter 3
Ion Plating System

The basic features of a simple ion plating unit are shown in Figure 24. There are four main units as follows:

(1) The vacuum and gas control unit.
(2) The gas discharge chamber.
(3) Evaporation sources.
(4) The high voltage electrode.

3.1 THE VACUUM AND GAS CONTROL UNIT

The reproducibility of ion plated coatings requires an operating environment which interferes as little as possible with the process of film formation. This involves the reduction of contaminant level by careful design of the vacuum system. Oil vapour and hydrocarbons emitted from gaskets and pumps can be degraded, contaminating the deposited coating (Holland, 1974b). Therefore metal gaskets can be used for sealing infrequently dismantled components but an elastomer seal is usually necessary on the vessel closure plate for ease of operation. In general, the vacuum system associated with the ion plating process employs two different principles to reduce the pressure in the vacuum chamber (Maissel and Glang, 1970). The first involves the physical removal of gasses from the vessel by exhausting the gas load to the atmosphere, by use of mechanical and vapour stream pumps. The other method of evacuation relies on condensation or trapping of gas molecules on some part of the inner surfaces of the chamber without discharging the gas by using ion pumps or cryo-pumps. The flow of large quantities of inert gas at relatively high pressures through the ion plating system does, however, impose considerable strain on most pumps. Diffusion pumps are probably the most widely used type for ion plating, and provided care is taken to prevent significant backstreaming of the pump fluid, their operation is quite satisfactory (Figure 25) (Maissel and Glang, 1970).

In ion plating processes, suitable pressures can be achieved by using mechanical pumps to rough out the vacuum chamber from atmospheric pressures down to about 10 Pa and then an oil diffusion pump with a liquid nitrogen trap to reduce the pressure in a short time to 10^{-4} Pa. This operation should reduce the partial pressure of reactive gases such as oxygen and water vapour to around 0.5×10^{-4} Pa (Goode and Burden, 1979). In industrial ion plating units a silicon fluid type (DC 704) is usually used in diffusion pumps for safety and durability.

Figure 24. Typical ion plating rig

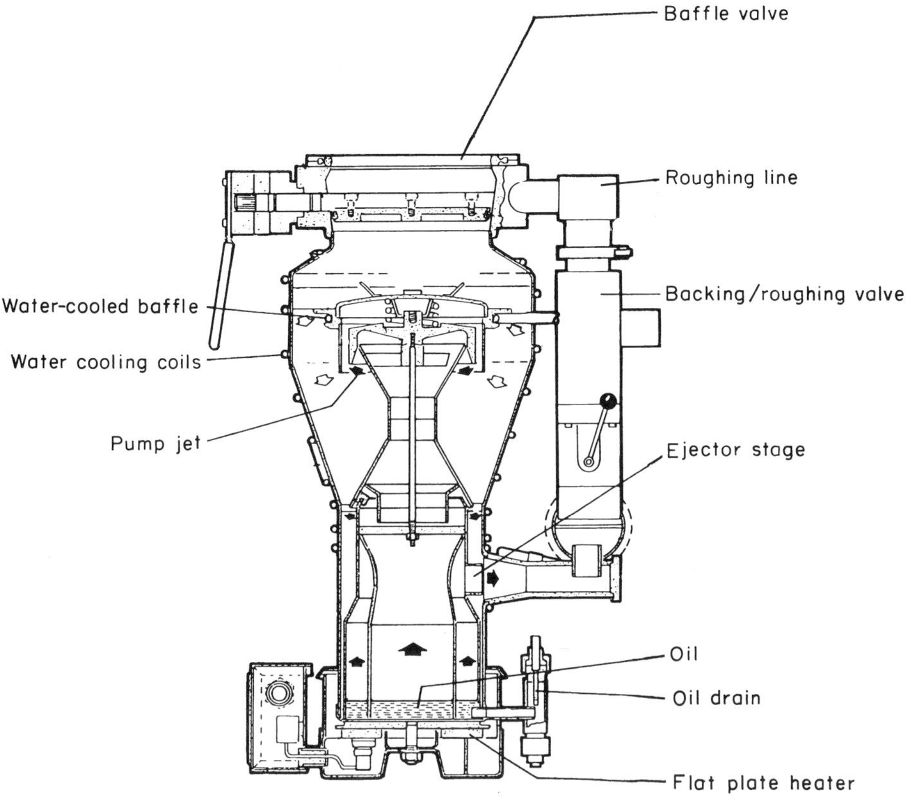

Figure 25. Schematic diagram of MK2 oil diffusion pump. (*Reproduced by permission of Edwards High Vacuum International*)

To avoid the problem of building up of contaminants during the ion plating process, the isolation valve is used as a throttle valve in order to allow the system to be continuously flushed with the discharge gas whilst maintaining a constant pressure inside the chamber. However, this may pose a real problem since if the throughput is too low, the system will be unable to flush contaminants out of the processing chamber. If the throughput is too high the pumps may operate inefficiently and backstreaming can occur.

In the case of reactive ion plating, a titanium sublimation pump could be attached to the coating chamber. The advantage of this pump is that although it does not pump out inert gases, it has substantial pumping rates for reactive gases such as oxygen, hydrogen, nitrogen, etc. (Weston, 1984).

The vacuum chamber and other parts of the vacuum system should be made from a material which exhibits minimum outgassing under heating and discharge conditions. For low vacuum operation, a mild steel chamber with a leak rate of better than 1×10^{-3} Pa litre s^{-1} is usually used, whereas stainless steel is used for a chamber operating at high vacuum. In systems which are made of stainless steel with polished walls and a liquid nitrogen trap, the pump speed-to-volume ratio provides about a decade lower pressure compared to a mild steel chamber at fixed pumping time (Forant and Habdanian, 1978). The principal problem in maintaining pumping system performance, however, is the accumulation of coatings on the fixtures and chamber walls. If the pump speed is virtually constant then the system performance variations can be attributed to increased gas loads either from outgassing or leakage. There is, however, less improvement possible by increasing pumping speed than by reducing gas loads. Significant shortening of pumping cycles can be achieved by careful leak correction and reduction of outgassing by proper preparation of the product, cleaning of the system and selection of low vapour pressure lubricants.

Gases used in the usual plating process are mainly dry inert gases such as argon or gas mixtures and are 99.99% pure. In reactive ion plating it is usually necessary to have a few per cent of reactive gas and the balance inert gas. Relative composition of gas mixtures can be very important to the product characteristics. The gas flow rate can be monitored using a suitable flow meter and the flow rate can be controlled using a fine needle valve or a mass flow controller combined with magnetic isolation valve. The valve can be set manually or can be controlled by a servo system from a gauge in the discharge system or foreline. It should be emphasized at this point that water vapour or condensation should be avoided by flushing the chamber with dry nitrogen or dry air and warming up the cooling lines by circulating hot water through them before letting up the system to air after the deposition cycle is completed.

3.2 THE DISCHARGE CHAMBER

The conventional ion plating process is carried out in a gaseous discharge and it is essential that the whole substrate 'sees' the same discharge conditions. It is therefore of prime importance to have a uniform environment to achieve uniform electric field distribution, uniform plasma, uniform deposition and uniform temperature distribution on the substrate. These are important factors which govern the quality of the required coating. Therefore, it is a necessity to construct a symmetric system in order to obtain a uniform electric field around the substrate. If the system is not symmetric, the electric field will be distorted and there will be a plasma density variation due to wall losses. The substrate holder should be surrounded by a ground shield to restrict ion bombardment and sputtering to the substrate only, otherwise the sputtered holder material will contaminate the growing film on the substrate. In order to prevent ion bombardment of the protected regions, the space between the target and the ground shield must be less than the thickness of the dark space. This means a gap of a few millimetres depending on the gas pressure and bias. However, since the thickness of the dark space decreases as the pressure increases, the gap width should be set for the upper pressure limit for operating the system. It is also important to ground other surfaces exposed to the plasma and keep them at a distance 5–10 times the cathode dark space length from the substrate surface (Mattox, 1982b). Gas pressure gradients must be avoided in the vicinity of the substrate since this causes discharge inhomogeneity.

The uniformity of the discharge in the vacuum chamber is strongly influenced by the position of the gas inlet in the chamber. The most uniformly distributed glow discharge corresponds to the best gas flow characteristic (Forant and Habdanian, 1978). This can be achieved by using the base plate inlet with an extension tube reaching the cathode area or using the top plate inlet, and in both cases the gas inlet should be placed opposite to the pumping port. In reactive ion plating a ring gas inlet can be used around the substrate to ensure better reaction. Fresh inert or reactive gases should be constantly admitted to and removed from the chamber during ion plating. The steady passage of relatively large amounts of fresh high purity gas through the system serves mainly to dilute the concentration of outgassed impurities present in the gas and it helps cooling of the substrate and other parts of the chamber. However, in the case of reactive ion plating, it is a prior condition to have an adequate supply of reactants which can undergo collisions and react to form the desired compound coating. This will lead to the consumption of the reactive gas and care must be taken to continuously replenish the consumed gas. This is frequently carried through a gas inlet valve controlled using the pressure measured in the chamber. Furthermore, in reactive ion plating there should be no stagnation zone in the gas flow and gases should be premixed before being injected into the chamber to ensure compositional uniformity. This is important in the case of titanium nitriding where colour variation which is related to nitride properties is seen to be very sensitive to pressure and gas mixing.

If the gases are admitted separately and their pressures are deduced from pressure values read in the main chamber, their actual ratio will be a function of the order in which they were admitted into the chamber. This is because in the pressure range used for ion plating the pumping rate in most vacuum systems is very sensitive to pressure. In order to prevent contaminating the substrate surface during sputter cleaning and the initial melting of the evaporated coating material, a shutter should be used. When the shutter is opened it should be kept well away from the plasma region and the evaporation source in order not to disturb the discharge or block the evaporated material.

3.3 EVAPORATION SOURCES

The developments in ion plating apparatus have led to new designs of vapour sources to expand the range of materials which can be deposited and to allow more flexibility in deposition rates so that process parameters may be optimized. However, each evaporation source has its advantages and disadvantages and the evaporation source has to be selected to meet the requirement for the end-product in terms of evaporation temperature, evaporation rate, coating thickness, coating purity, and economic considerations. Therefore, depending on the type of ion plating to be performed and the desired properties of the ion plating parts, a broad range of evaporation sources are available and are discussed below.

3.3(a) Resistance Heated Sources

These are the most commonly used vapour sources with the evaporation material placed in a filament, strip, boat or crucible and heated by passing a current through it. The type and the shape of the source depend on the type of the evaporated material, its evaporation or sublimation temperature and the coating thickness required (Bunshah, 1982). The sources could be made of tungsten, molybdenum, tantalum or carbon. In other cases ceramic crucibles or inserts such as alumina, boron nitride, beryllium oxide or boron nitride/titanium diboride (BN/TiB_2) are also used. When the material is evaporated from the source it may react with the boat material causing deterioration or may contaminate the growing film on the substrate. The most widely used sources are intermetallic boron nitride/titanium diboride boats. These types of boats are commercially used to hold evaporant materials because they are comparatively sturdy, inexpensive, reliable and simple to operate (Glang *et al.*, 1970). The intermetallic sources which are shown in Figure 26 have been developed over the past 25 years particularly for the metallizing industry (Parent, 1978). The boats have an electrical resistivity within the range

Figure 26. Intermetallic boats. (*Reproduced by permission of Borax Consolidated (Borides) Ltd*)

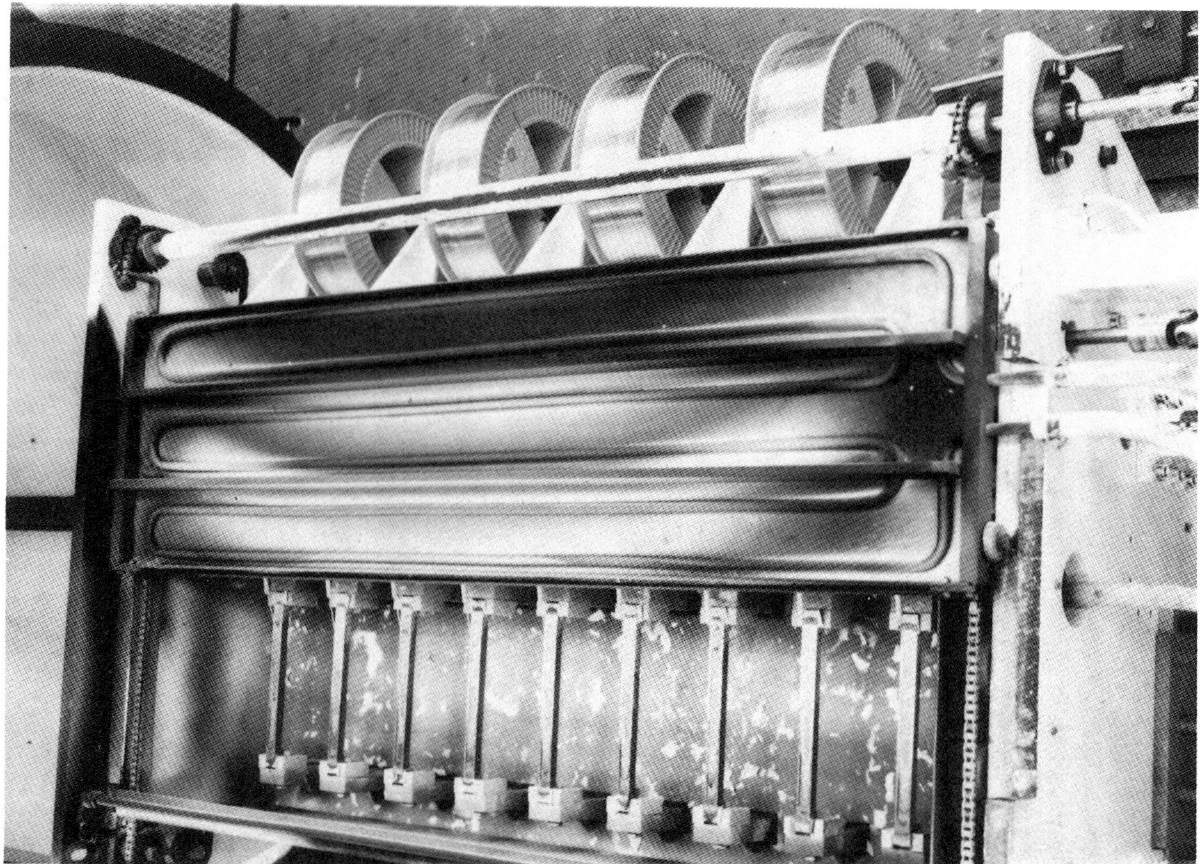

Figure 27. Wire feeder for continuous evaporation of metal at high evaporation rates (from Ahmed, 1986a). (*Reproduced by permission of The Institute of Metals*)

250–1000 micro-ohm cm and have been designed to be stable in the vacuum and corrosive environment. The boats are mainly used with wire feed or slug feed to continuously deposit thicker coatings of low melting point materials such as aluminium. Figure 27 shows a wire feed system designed to continuously evaporate aluminium wire from a boron nitride/titanium diboride boat in an industrial ion plating coating unit. The wire is fed at a certain speed to match the evaporation rate from the boats. Figure 28 shows a warning device attached to the wire feed system to warn the operator in the case of the wire running out or snapping due to high dragging torque during the process (Ahmed, 1984). During operation a hot evaporator boat loses heat by three means (Parent, 1978):

(1) By supplying latent heat to the material being evaporated.
(2) By radiation loss from the hot boat walls.
(3) By conduction loss to the water-cooled clamps.

In general, the latent heat can be estimated from the known latent heat of melting and evaporation and the specific heat of the metal once the rate of evaporation is known (Parent, 1978). Heat loss by radiation can be estimated from the Stefan-Boltzmann law assuming a crucible temperature of 1500°C and an emissivity of 0.62. The conduction loss to the water-cooled clamps can be estimated from the known thermal conductivity of the composite and an estimate of the temperature profile near the clamps. It has been estimated that metal melting and evaporation consume 15–17% of the input power. Radiation loss consumes 45–55% and conduction loss to the water-cooled clamps consume 30–35% of the power used (Parent, 1978). The water-cooling loss can be minimized by the use of graphite end pieces or spring contact type clamps. The largest loss of

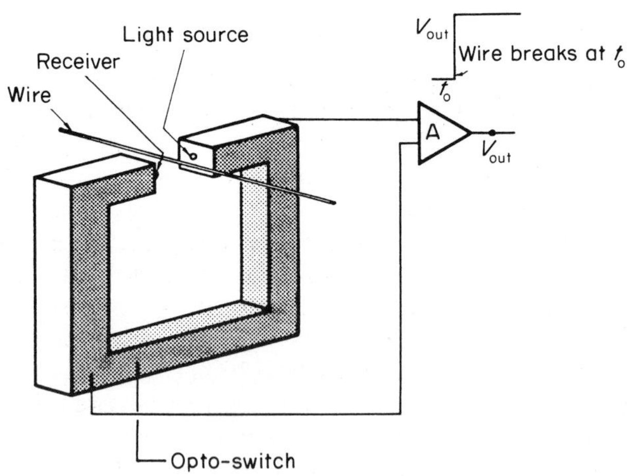

Figure 28. Wire feeder warning device (from Ahmed, 1984). (*Reproduced with permission from Vacuum vol. 34,* ©*1984 Pergamon Journals Ltd*)

power is in radiation. Although some radiation loss may be useful in improving the vacuum pump down time by releasing absorbed gases on the chamber walls, it may, however, cause excessive rise in the substrate temperature and result in bad adhesion or substrate damage in the case of plastic. Radiation loss can be kept to a minimum by designing the boat with a maximum ratio of evaporation area to total area (Hammer, 1970), or by surrounding the boat with a radiation shield (Parent, 1978). The use of the latter may save dissipated power, reduce radiation damage, equalize boat temperature and can be designed to direct the vapour cloud to the substrate in a more efficient manner during ion plating of special articles. The radiation shield can be made of metal such as molybdenum or insulator such as graphite. The lifetime of the boat used is usually limited by several factors including operating time, working temperature, initial conditioning of the boat, thermal shocks (when the current increases suddenly) and reaction between the molton metal and the boat material. Therefore, in order to increase the working life of a boat it is important to follow the manufacturer's operating procedure. However, some problems may be encountered during the operation of a boat such as boat cracking or spitting of the evaporant. Cracking of the boat may result from a tightly clamped boat or as a result of a thermal shock. A high moisture level (owing to careless storage) in the boats may also increase the probability of cracking during the first heating cycle (Borax Consolidated, 1985). Spitting can result in pinholes in the coating and may be caused by un-isolated wire feed, incorrect feed rate and fluctuating boat temperature. It can also occur when there is a severe erosion groove in the cavity, resulting in a deep pool of aluminium evaporant, for example. It is important to isolate electrically the wire feed from the boat, otherwise electrical discharge may take place, leading to spitting.

3.3(b) Flash Evaporation

The range of materials which have been ion plated with intermetallic resistance heated evaporation boats is rather limited and is confined to the elemental metals with melting temperatures of less than 1300°C, usually aluminium. The technique also has limitations when evaporating alloys or compounds due to fractionization of these evaporating alloys or components. Also some evaporating materials may alloy or react with the boat and it is not possible to evaporate high temperature materials such as refractories. To overcome some of the above difficulties and expand the range of materials which can be evaporated by resistance heating, several new approaches have been developed such as flash evaporation (Bunshah, 1982). In this technique (shown in Figure 29) the material to be deposited is usually in powdered form and is fed continuously into a preheated boat. The temperature of the boat is set as high as necessary to evaporate instantaneously the least volatile component of the material being used. The constituents of the powder, when coming into contact with the boat, are vaporized instantly, thus preventing any functional or partial decomposition of the material. No material should accumulate in the boat, and

Figure 29. Flash evaporation mechanisms. (*Reproduced by permission of Balzers AG*)

the vapour which is produced from a uniform powder feed should generally provide a surface coating with the same composition as the original material. Parameters to be controlled in flash evaporation are particle size, rate of powder delivery and preheat temperature of the reaction boat. By using flash evaporation, Spalvins achieved successful ion plating of binary alloys of (Pb–Sn) and (Cu–Au) retaining the original stoichiometry (Spalvins, 1970).

3.3(c) Electron Beam Evaporation

Probably the most important advance in ion plating sources was that developed by Chambers and Carmichael (1971) who, by using a double chamber construction and differential pumping, were able to operate a hot filament electron beam gun at ion plating pressures. This type of system which is shown in Figure 30(a and b) has very significantly increased the potentials of ion plating and makes possible high deposition rates and evaporation of virtually all metals

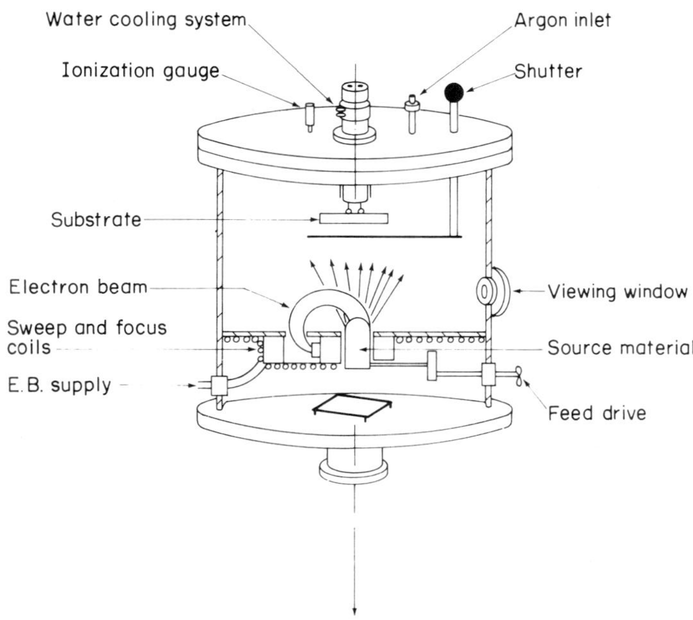

(a)

(b)

Figure 30. (a) and (b) Electron beam evaporation. (*Reproduced by permission of Ferrofluidics Ltd, UK*)

including molybdenum and tungsten with melting temperatures of up to 3500°C. Evaporation is generally confined to water-cooled copper crucibles which have to be used for the evaporation of reactive or refractory metals. Electrons are emitted from the hidden hot filament and accelerated to high energies in the electric field, then deflected magnetically by 270° and thus focused on the crucible. The voltage required to accelerate the electrons towards the crucible is in the range of 6–10 kV. Power densities as high as 10 kW cm^{-2} can be achieved with this type of electron gun. Different crucible sizes and shapes can be used with this gun. However, the disadvantage of the hot filament electron beam gun is that it should be operated at high vacuum, whereas the ion plating process is carried out at lower vacuum. Therefore, systems utilizing an electron beam gun must have a differential pumping aperture to separate the plasma area from the higher vacuum of the electron beam area. The other type of electron gun is the hollow cathode electron beam gun (Dugdale, 1966) which consists of a hollow cylinder or a sphere made of a refractory metal since it operates at elevated temperatures. Figure 31 shows a schematic diagram of the

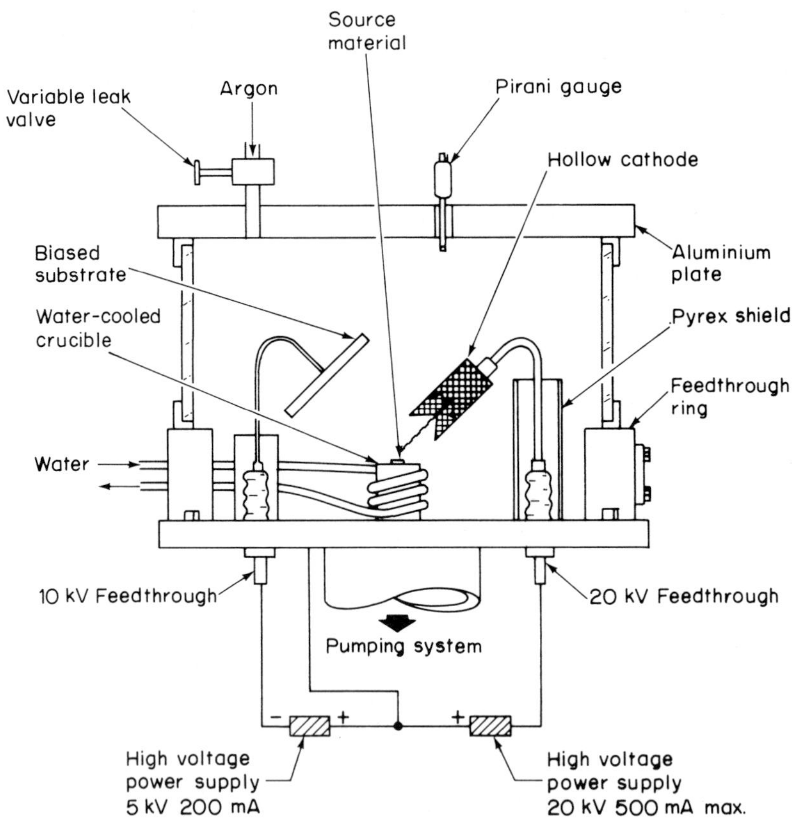

Figure 31. Schematic representation of a hollow cathode gun (from Wan *et al.*, 1971). (*Reproduced by permission of the American Institute of Physics*)

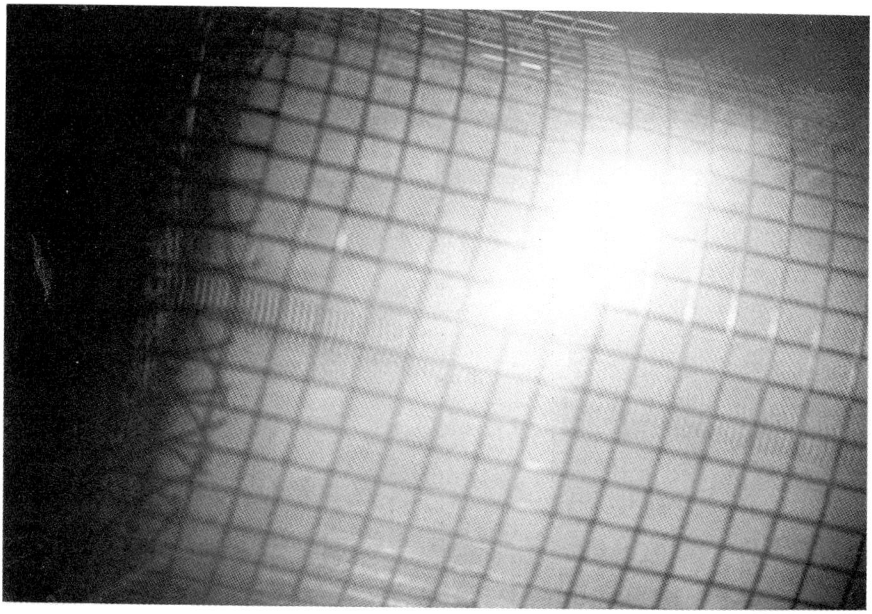

Figure 32. Hollow cathode discharge inside a perforated cylinder

hollow cathode gun. The hollow cyclinder or sphere has an aperture in one end and is immersed in an argon atmosphere. The aperture provides a sufficient amount of gas inside the gun to sustain the discharge. A high voltage d.c. power supply is used and the gun is made as the cathode of an electrical circuit with the anode connected to the crucible. As soon as the glow discharge is generated, an electron beam is emitted in a conical shape from the aperture (Figure 32). The main advantages of the hollow cathode guns are that they can operate at glow discharge pressures, do not use a hot filament and are easy to design and construct. However, the primary disadvantage is that the gun is also sputtered in the process, which contaminates the coatings, so that shielding is required to prevent the sputtered atoms of the gun from depositing on the substrate. Furthermore, the mode of operation is greatly influenced by gas pressure. At high pressures, the gun operates at the hot cathode mode while at lower pressures the gun operates at the cold cathode mode which is very sensitive to pressure variation (Dugdale, 1966).

Figure 33 shows the current–voltage characteristics of a simple hollow cathode gun at different discharge pressures. By adjusting the pressure and the potential applied to the gun, maximum electron beam power can be achieved.

3.3(d) R.F. Induction Evaporation

This evaporation source consists essentially of a ceramic crucible such as alumina or boron nitride/titanium diboride with the plating metal or alloy inductively

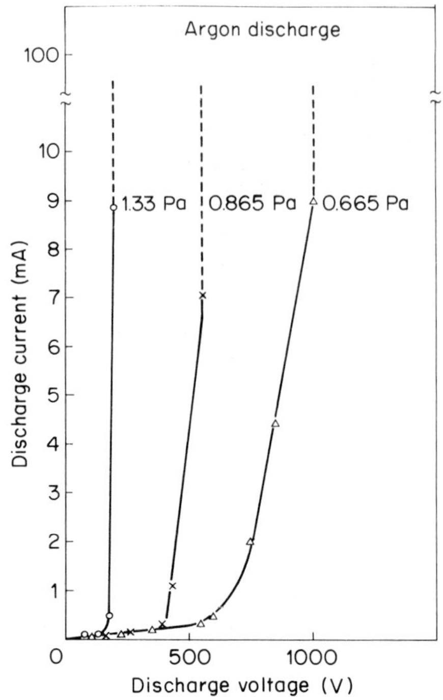

Figure 33. *I–V* characteristics of a hollow cathode discharge

evaporated as shown in Figure 34. For evaporating reactive metals such as Be or Ti a water-cooled copper crucible may be used (Bunshah, 1982). The induction coil can be mounted in the glow discharge without producing an arc if the operating frequency is reduced to about 75 kHz. This can be accomplished by using an induction generator reduced to this level from the more normal

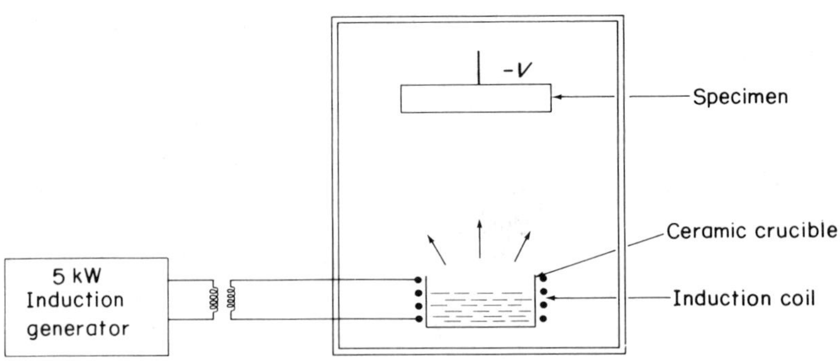

Figure 34. Schematic representation of an r.f. induction evaporation

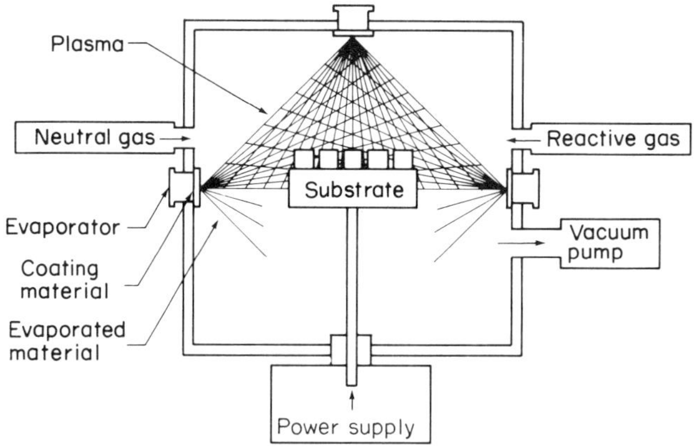

Figure 35. Schematic representation of arc evaporation (from Hatto, 1986). (*Reprinted with permission from Vacuum Vol. 36, ©1986 Pergamon Journals Ltd*)

metallurgical operating frequency of about 450 kHz (Spalvins and Brainard, 1976). The induction heating source usually increases the ionization efficiency of the evaporation flux by enhancing the discharge.

3.3(e) Arc Evaporation

Vacuum arc melting processes have been used for a long time to melt and refine active metals such as Al, Ti and Hf. In this method, a high current electrical discharge (arc) is used as a heat source to evaporate different materials. In recent years, arc evaporation techniques have been used commercially for reactive ion plating of titanium nitride (Figure 35) (Hatto and Teer, 1986). In this case, the electrode material consists of pure titanium. The arc is supported in the generated flux and can therefore be maintained under high vacuum conditions, unlike the glow discharge used for sputtering and ion plating. The arc can be operated in a variety of modes (Dorodnov and Petrosov, 1981), that usually used being the discrete spot mode where the arc consists of one or more discrete spots moving randomly on an electrode. Figure 36 shows an arc spot on titanium cathode. Because of the very high energy density (10^{11} W m^{-2}) when the arc is operated in this mode, electrode material in the vicinity of the spot passes directly into the vapour phase without the electrode undergoing bulk melting. Thus the arc source may be used to produce a vapour flux in any orientation, i.e. evaporation may take place sideways, downwards or upwards and thus sources may be located in any desired position within a coating system (Hatto and Teer, 1986). Figure 37 shows the gaseous glow discharge characteristics where a vacuum arc is formed immediately after the abnormal discharge. As can be seen from Figure 37 the arc is a low voltage, high current discharge which makes it particularly attractive from an electrical isolation/safety point of view

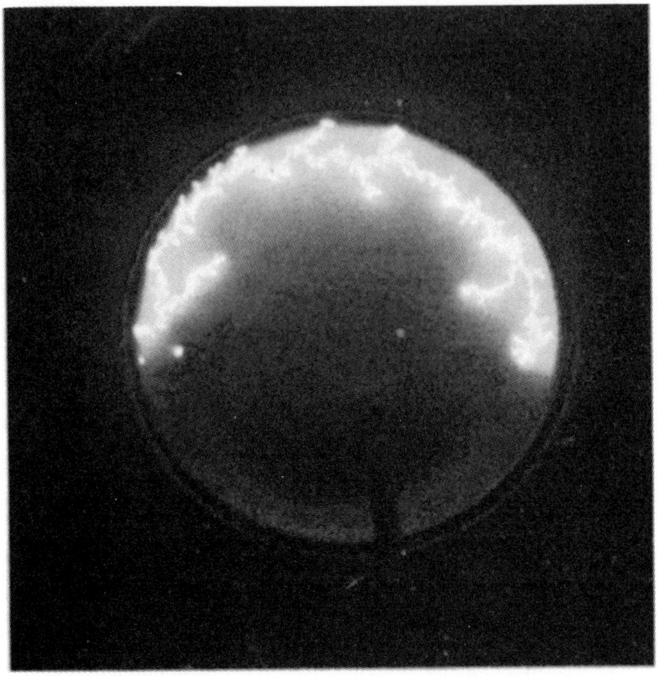

Figure 36. Vacuum arc spot on titanium cathode (from Hatto, 1986). (*Reprinted with permission from Vacuum vol. 36, ©1986 Pergamon Journals Ltd*)

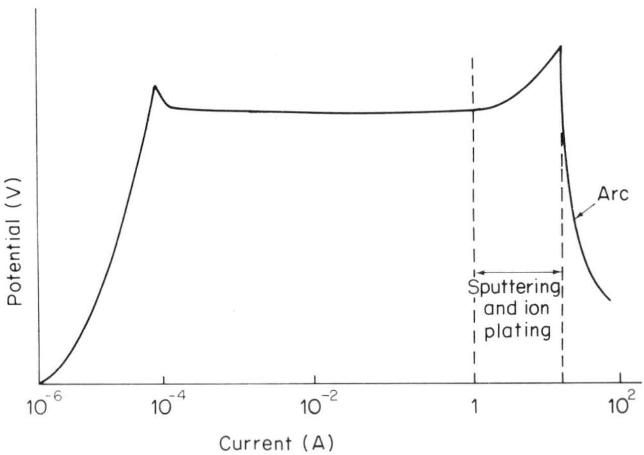

Figure 37. The characteristics of arc glow discharge (from Hatto, 1986). (*Reprinted with permission from Vacuum vol. 36, ©1986 Pergamon Journals Ltd*)

(Hatto, 1986). Typical operating conditions are 15–30 volts d.c. and 50–150 amps; a.c. operation of vacuum arc sources has also been achieved and appears to show certain advantages over the d.c. mode (Dorodnov and Miroshkin, 1981). The evaporated materials ejected as a result of the arc discharge contains atomistic material, which is both ionized and neutral, and various sized clusters of atoms.

3.3(f) Sputter Ion Plating

Another source for depositing materials in an ion plating system is the use of sputtering targets (Schiller *et al.*, 1977; Ridge *et al.*, 1981; Dugdale, 1977). Ion plating with a sputtering target (Figure 38) as a coating source is very similar to bias sputtering with the exception that higher bias voltages are used on the substrate as in conventional ion plating. However, high substrate voltages (2000–5000 V d.c.) that are commonly used in ion plating cannot be applied to sputter ion plating. Furthermore, it is necessary sometimes to continuously decrease the potential on the substrate so that at some point the deposition rate becomes greater than the normal resputtering rate. This is particularly true when depositing an easily sputtered material such as gold. In order to increase the sputtering rate a d.c. or r.f. magnetron sputtering target can be utilized. In this

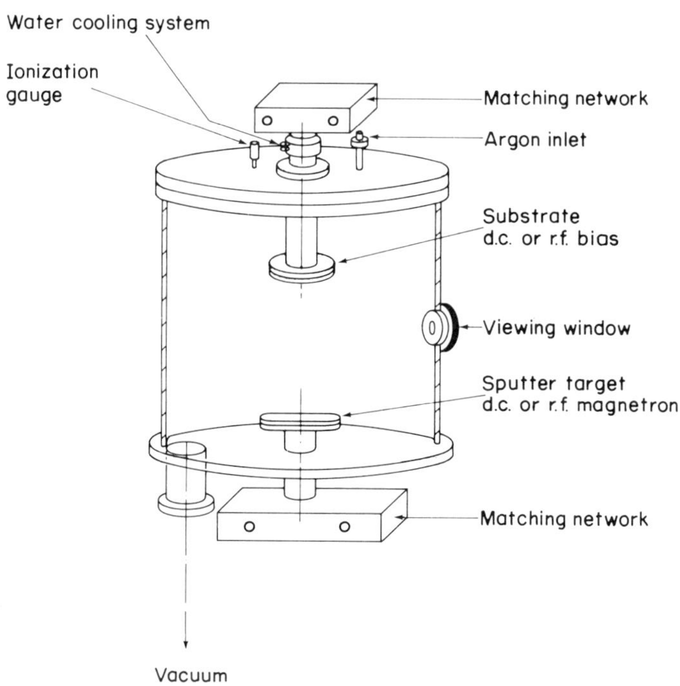

Figure 38. Schematic diagram of sputter ion plating

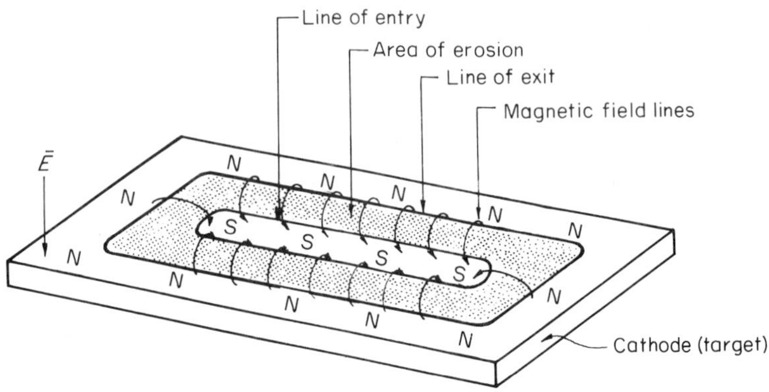

Figure 39. Sputtering magnetron target

system (Figure 39) a magnetic field is used to concert with the cathode (target) surface to form electron traps which are so configured that the **E** × **B** electron drift currents form closed paths (Thornton, 1978). Thc magnetic field is generated by the inclusion of permanent magnets or electromagnets directly behind the target. The magnets are arranged so that there is at least one region in front of the cathode surface where the field lines form a closed path. Furthermore, the interaction between the electric and magnetic fields forces the electrons within the plasma to travel in long helical paths in very close proximity to the target surface, thus creating more collisions and ionizations of the argon gas atoms. The resulting ionized plasma acts in a positive space charge mode with a current density to the target of some two orders of magnitude greater than in conventional d.c. or r.f. diode sputtering (Thornton, 1979). The target is usually cooled by circulating water through it. Such cooling is important since excessive heating due to ion bombardment may crack or disintegrate the target. R.f. bias is usually used when sputtering dielectric targets. When r.f. bias is applied to such targets, it becomes the anode during the positive half of the cycle, hence attracting electrons, and the ion current is usually zero. During the negative half of the cycle, the target becomes the cathode and the ion current is maximized when the peak potential is applied. As the negative potential is reduced the more mobile electrons can again cross the collapsing plasma sheath, thus offsetting the ion current until the net current reduces to zero. By using an alternating potential of a suitable frequency, the excess electron current will be accumulated on the surface of the dielectric target and result in a negative self bias (Figure 40). The negative d.c. bias achieved is sufficient to equalize the ion and electron currents and ensures that a continuous sputtering action occurs throughout the complete r.f. cycle. Typical r.f. frequency used is 13.56 MHz and the power density on the target may be up to 120 W cm^{-2} (Chapman, 1980). When sputtering a dielectric material, an r.f. bias is also applied to the substrate to achieve better stoichiometry and adhesion. However,

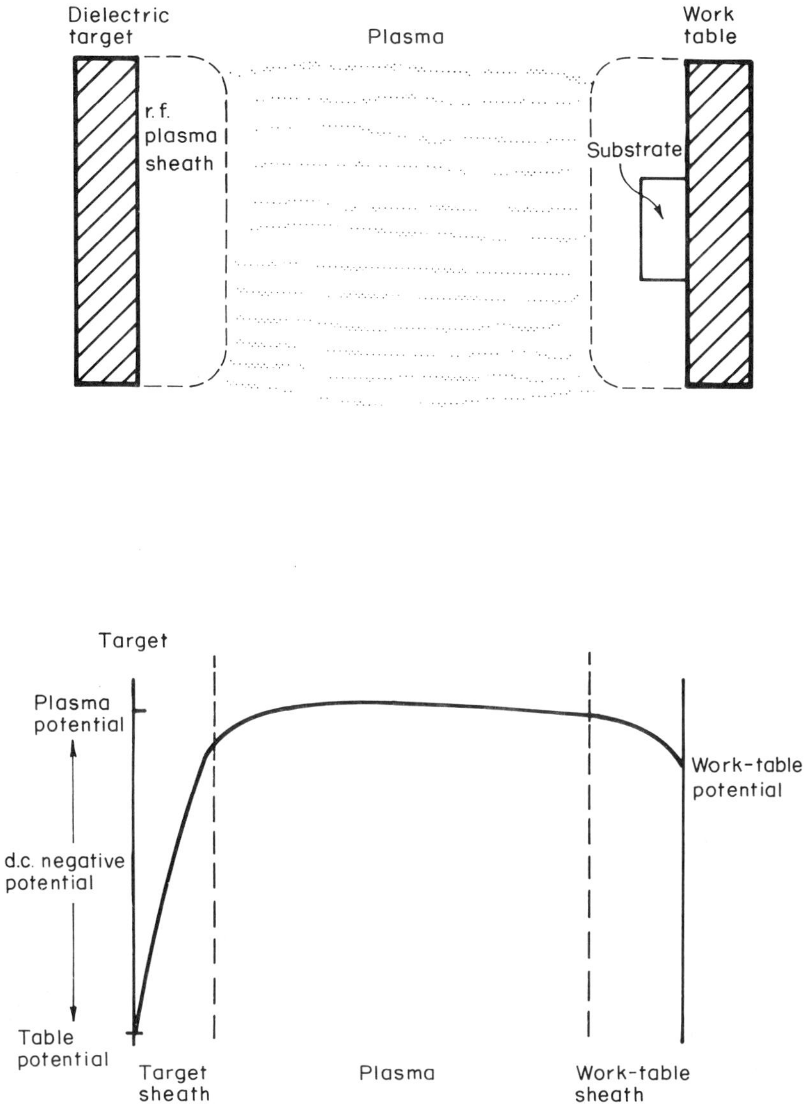

Figure 40. Negative potential distribution in an r.f. sputtering system

the sputtering source has a limited lifetime and this technique is mainly suitable for coatings up to a few microns in thickness. This technique can be used to deposit alloy coatings, such as Co–Cr–Al–Y alloys for the protection of turbine blades (Coad and Restall, 1982).

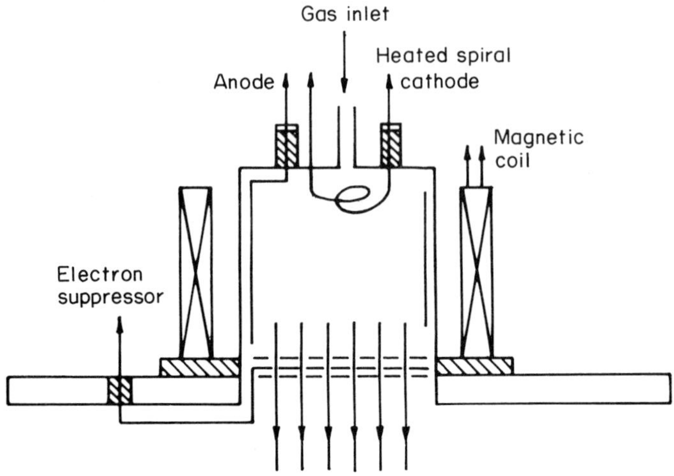

Figure 41. Schematic diagram of ion beam source

3.3(g) Ion Beam Evaporation

In this method, a Kaufman broad beam ion gun source is used to sputter clean the substrate and to sputter target material for deposition (Weissmantel, 1976). The source consists of a hot filament for emitting electrons (cathode) and a molybdenum cylinder (anode) which is surrounding the cathode and providing a radial electric field (Figure 41). The cathode and anode are installed inside a cylindrical discharge chamber which contains a gas inlet and extraction grids. A solenoid electromagnet is mounted on the outside of the ion source chamber and is producing an axial magnetic field, perpendicular to the electric field in order to deflect the electrons into cycloidal paths (Kaufman, 1963). The ion source is mounted in a vacuum chamber capable of maintaining the pressure inside the source to about 0.01 Pa at the operating conditions. During operation inert gas (usually argon) is admitted to the discharge chamber and is partially ionized by low energy electrons accelerated from a thermionic cathode filament. The magnetic field enhances the ionization of the gas by increasing the electron paths and hence increases the ionization due to multiple collisions. A low voltage (500–1000 V) is applied to the anode to accelerate the ions and therefore to gain an energy corresponding to the anode potential. The ions are then accelerated towards the extraction grids then extracted as a collimated ion beam. The extraction grids consist of a screen grid which is held at cathode potential and the acceleration grid which is held at a small negative potential to add to the electric field for extraction of ions as well as to provide an electrical barrier to prevent backstreaming of electrons from the beam region into the discharge plasma (Kaufman, 1974; 1978). After extraction the ions bombard the target at high energies. Deposition atoms are obtained by sputtering the target with a portion of the ion beam while the other portion bombards the substrate during the growth of the film. In order to achieve film growth, the deposition rate on the substrate should be higher than the resputtering rate of the film material. This technique has a distinct advantage that the ion bombardment parameters can be controlled independently of other deposition parameters. However, the disadvantages are the low deposition rate as compared to resistance heated crucibles, and the bad throwing power since the coating is virtually carried out at high vacuum (i.e. no scattering due to the negligible number of gas atoms). Ion beam guns can produce high current densities up to 1 mA cm^{-2} with ion energies up to 2 keV.

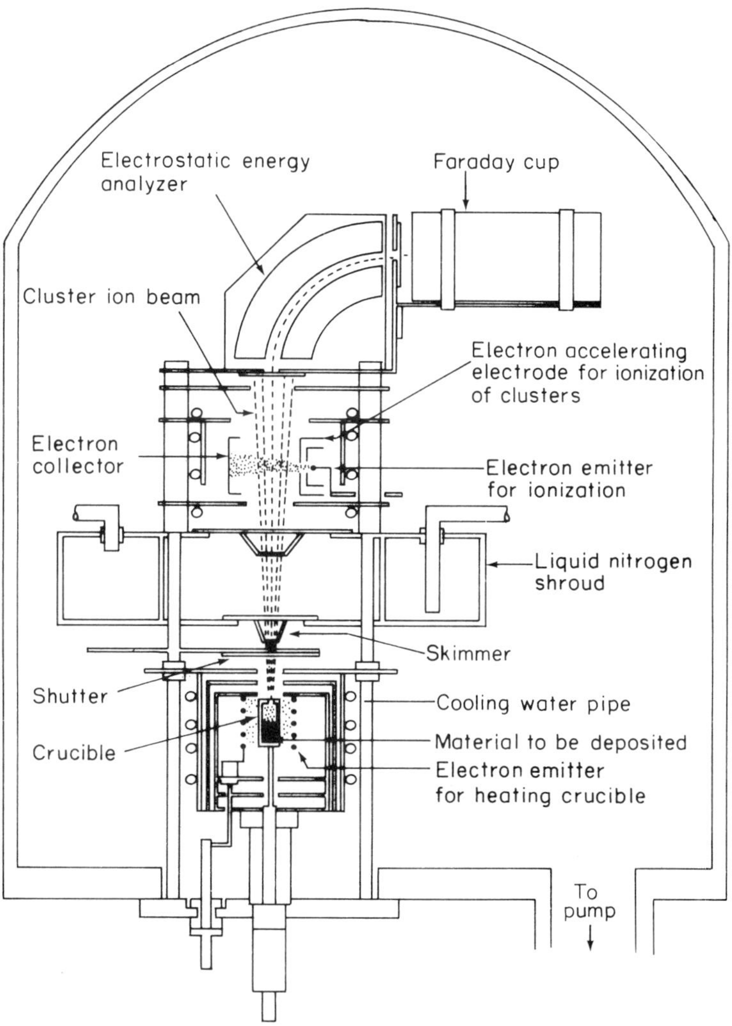

Figure 42. Schematic diagram of ionized cluster beam deposition (from Yamada and Takagi, 1981). (*Reproduced by permission of Elsevier Sequoia SA*)

3.3(h) Ionized Cluster Beam Deposition

Ionized cluster beam deposition technique was originally developed by Takagi *et al.* (1976). In this technique (Figure 42), a vapour flux is generated inside an evaporation cylindrical nozzle source designed to cause condensation of the vapour into clusters of several hundreds of atoms as they exit the cell (Takagi, 1986). The clusters are formed by supersaturation caused by adiabatic expansion through the nozzle without the mixture of any inert gas. The clusters pass through a separate ionization region and are ionized by electron bombardment. The ionized clusters are then accelerated towards the negatively biased substrate and strike it at energies of several eV per atom and disperse on the surface. Therefore, this technique combines a high evaporation rate with some of the benefits of ion bombardment. The advantages of this technique are the elimination of problems caused by the space charge effects due to the extremely small charge-to-mass ratio as well as the ability to control the cluster energy by controlling the substrate bias. The cluster size can also be controlled to some degree by controlling the pressure difference across the exit nozzle and by selecting a suitable nozzle shape. Coatings prepared by this technique have been reported to be highly smooth due to enhanced surface and atom migration which increases by increasing the acceleration voltage (Yamada and Takagi, 1981). One of the limitations of this technique is the poor throwing power due to the absence of gas scattering, which lands to difficulty in coating complex articles with hidden areas.

3.4 THE HIGH VOLTAGE ELECTRODE

As the ion plated substrate is usually held at high voltages, it is necessary to use a reliable electrode that can operate successfully at different ranges of voltages and temperatures. Thus a substrate holder with a high breakdown voltage can be designed by selecting a suitable insulator such as ceramic, PTFE or tefnol. The breakdown voltage is normally rated for breakdown in air (Maissel and Glang, 1970). Since most of the input power appears as substrate heating, it is also necessary to cool down the substrate and the holder by circulating cooling water through it at a flow rate which depends mainly on the temperature rise in the substrate. Substrate temperatures as high as 500°C can be reached without water cooling. Figure 43 shows schematically a typical water-cooled high voltage substrate holder. The holder is usually made in the form of copper rod. A grounded shield, also shown in the Figure, assists the suppression of the glow discharge between the biased holder and the earthed surroundings. The geometry of this shield has been discussed earlier in Section 3.2. Care should be taken to check the conductivity of the water used, to prevent any possibility of a breakdown. When an r.f. ion plating technique is used, the design of the substrate holder is slightly more complex than the d.c. holder. A typical problem of an r.f. system is that adequate grounding of the substrate assembly cannot be granted all the time. Therefore, a short circuit under d.c. conditions may

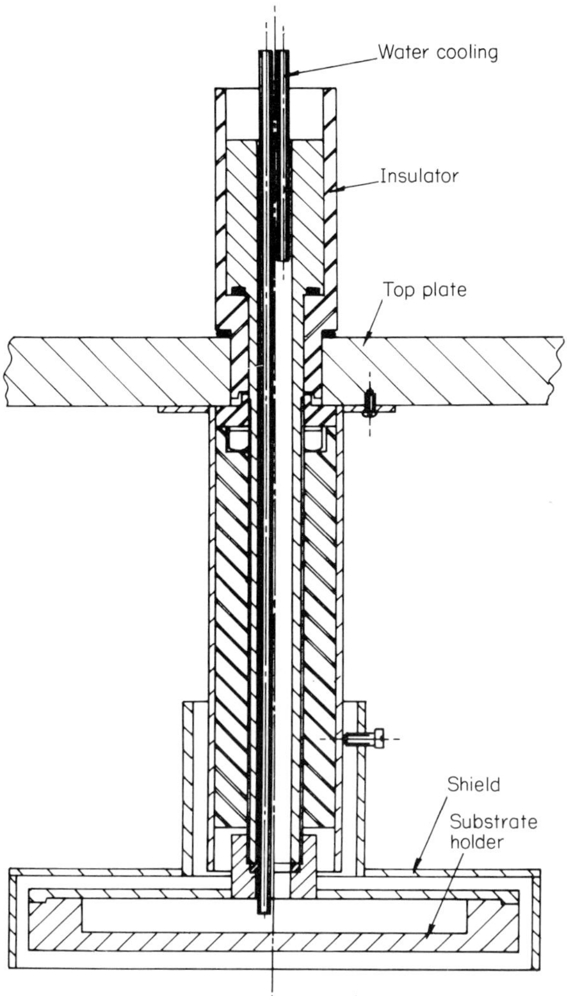

Figure 43. Water-cooled high voltage feedthrough

exhibit considerable r.f. impedance. If the substrate assembly is not adequately grounded, undesirable r.f. voltages can develop on its surface. Figure 44 shows a typical water-cooled r.f. electrode. The electrode has an earthed shielding around the non-active surfaces which carry r.f. voltages. Earthed shielding used at pressures of up to 2 Pa is usually designed to prevent gas breakdown by keeping the separation of the shielding from the r.f. electrode to less than 6 mm (Nordiko Ltd., 1986). Water-cooling pipes should be long enough and well grounded to prevent any possible electrical shock. Good earthing can be provided by selecting a true earth ground point and using a short, heavy copper strap with a large surface area. Other parts of the r.f. ion plating system should be

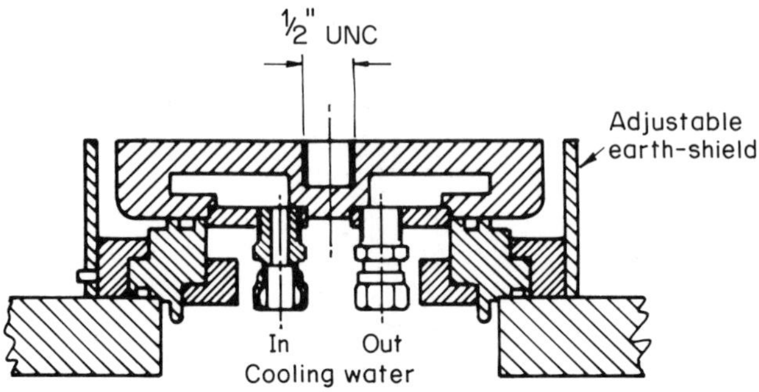

Figure 44. Water-cooled r.f. electrode. (*Reproduced by permission of Nordiko, Ltd*)

well grounded as well to protect the operator and equipment from r.f. burns and d.c. breakdown.

When coating a substrate of complex geometry such as that shown in Figure 45, several problems may arise due to the non-uniformity of the electric field surrounding the substrate. Through holes, blind holes and re-entrant corners can give a hollow cathode effect in which secondary electrons are trapped and focused, giving localized regions of high plasma density causing local heating and high sputter erosion. Corners and edges are high field regions where high sputtering rates are to be found. This may prevent these areas from forming a thick coating because of the high sputtering, or may give excessive erosion. Points or other high surface-to-volume geometries may become overheated because of poor thermal transfer properties. The author (Ahmed and Teer, 1981)

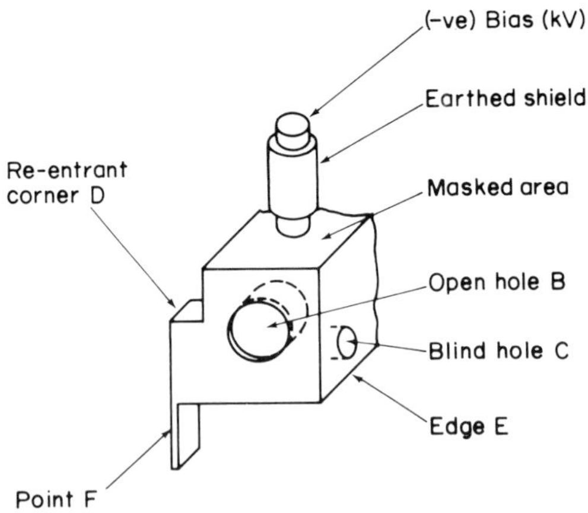

Figure 45. Geometrical effects in ion plated samples

has reported that although the presence of a hollow cathode discharge reduces coating thickness, it assists the formation of a highly adhesive dense structure (Figure 46). When depositing on an insulating substrate a d.c. or r.f. bias could be used. In the case of a d.c. bias, a high transmission negative grid can be placed in front of the substrate at a distance of around 1 cm from the substrate. Care should be taken to ensure that the substrate surface is not contaminated by material sputtered from the grid. For this reason it is always preferable to coat the grid with the evaporating material prior to coating the substrate. Other techniques used to ion plate insulating substrates include the deposition of a very thin metal film to make the surface conductive, followed by ion bombardment of the surface before continuing the deposition (Ahmed, 1985a).

3.4(a) Substrate Temperature Measurement

In certain applications it may be desirable to monitor the substrate temperature during ion plating to prevent the change in the mechanical properties of the substrate such as hardness and strength. Continuous monitoring of the substrate temperature during ion plating by methods such as optical pyrometry can give misleading results due to the effect of plasma discharge. An alternative technique for measuring the average substrate temperature is the use of Ni/Cr, Ni–Al thermocouple. The thermocouple which is shown in Figure 47 can be introduced into the vacuum chamber through a specially designed insulating feedthrough.

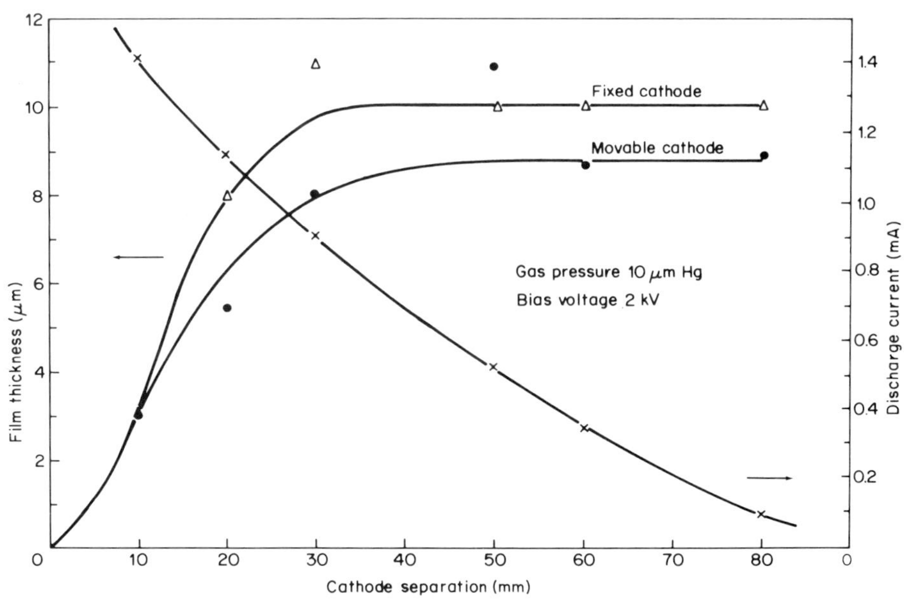

Figure 46. (a) Film thickness (△) and discharge current (×) dependence on cathode separation in a hollow cathode discharge (from Ahmed, 1981). (*Reproduced by permission of Elsevier Sequoia SA*)

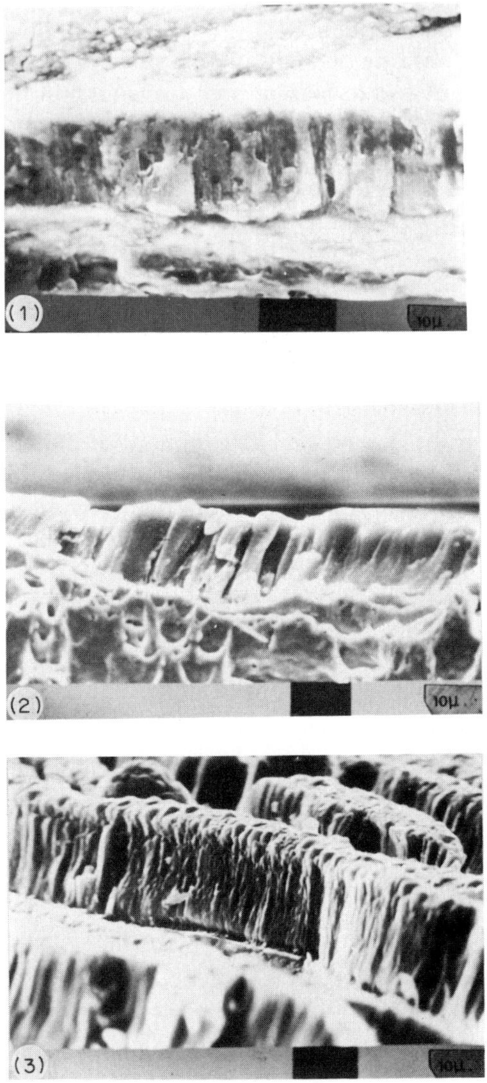

Figure 46. (b) The structure of ion plated aluminium in a hollow cathode discharge (from Ahmed, 1981). (*Reproduced by permission of Elsevier Sequoia SA*): (1) 2 kV, 10 μm Ar, 1 cm separation; (2) 2 kV, 10 μm Ar, 3 cm separation; (3) 2 kV, 10 μm Ar, 8 cm separation

The thermocouple wires are spot welded to the substrate and biased at substrate potential. Therefore, the wires and the temperature monitoring meter should be isolated from earth to prevent any damaging potential to the instrument. This can be done by inserting the thermocouple wires into a glass

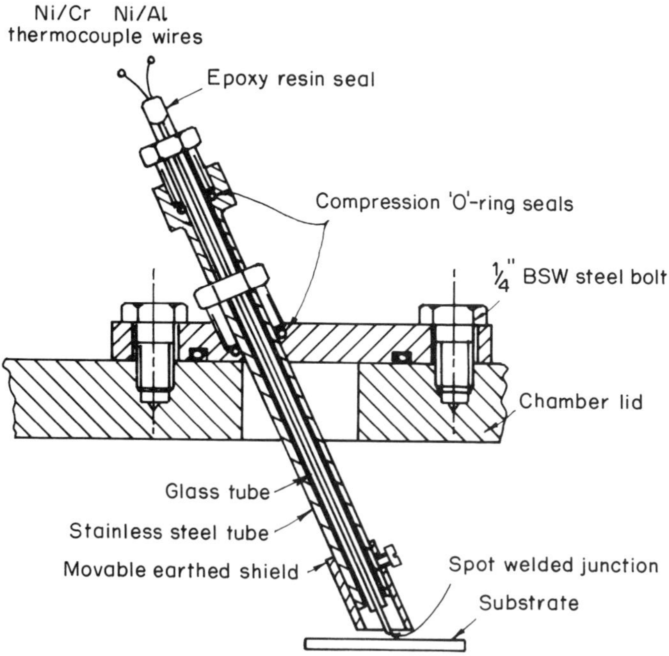

Figure 47. Thermocouple feedthrough attached to the high voltage biased substrate

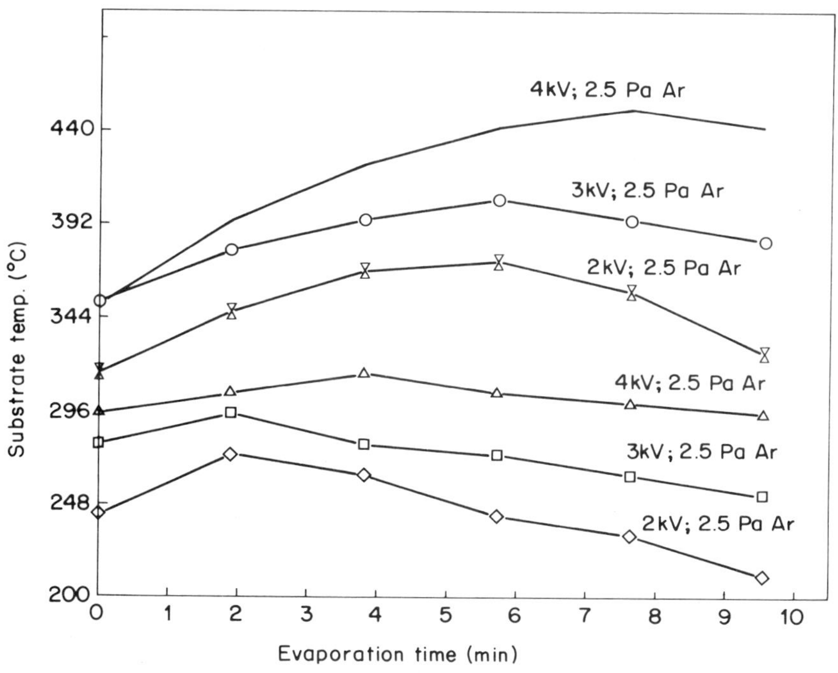

Figure 48. Temperature rise of the substrate at different ion plating conditions

tube which is consequently inserted into a stainless steel earthed tube. To protect the thermocouple wires from ion bombardment and consequent damage and also to avoid false readings, a moveable earthed shield is fitted to the end of the stainless steel tube to keep the glow discharge away from the spot welded area (Abu Zeid, 1982). By using this technique the substrate temperature can be monitored during sputter cleaning and coating. Figure 48 shows the relationship between the discharge current and substrate temperature at different discharge biases. It is evident that without water-cooling, average temperatures higher than 400°C can be reached at high substrate bias and high argon gas pressure. When monitoring the substrate temperature in an r.f. ion plating, electrical isolation of the thermocouple is more difficult since the thermoelectric e.m.f. of the couple should be preserved whilst removing both r.f. and d.c. offset components. The r.f. component may be removed by r.f. chokes whilst the d.c. component can be removed by using a thin insulator of mica or ceramic between the couple junction and the substrate platform or dummy substrate (Chapman, 1980).

3.4(b) Power Supply

H.V. d.c. power supplies used in conventional ion plating should be capable of supplying a negative voltage of 0–5 kV and 1 mA cm^{-2} current density to the cathode. Highly smoothed or regulated power supplies are not required. However, the output of the supply should have a good reactance to prevent overloading if an arc occurs in the system. Output cables should be capable of withstanding the maximum d.c. output voltage.

Commercial r.f. power supplies mainly use crystal controlled power amplifiers. Such power supplies have a capacitive output to provide several thousand volts peak-to-peak amplitude and operate at the allowed frequency of 13.56 MHz. The output impedance of the power supply is usually designed to be 50Ω to feed the power from the supply to the electrode. However, impedance matching between the power supply and the electrode, in order to get maximum power into the target or substrate, is often a problem. If the output impedance of the supply did not match the impedance of the cable, a proportion of the r.f. power output would be reflected back to the supply and cause damage by overheating the main output voltage (Mattox, 1982b). In general, r.f. power requirements needed to sputter clean insulated substrates or to sputter targets are typically 5–10 W cm^{-2} (Chapman, 1980).

Chapter 4
The Ion Plating Process

As we have just seen from the previous chapters the conventional ion plating unit is basically a diode system consisting of two electrodes. The specimen to be coated is attached to a water-cooled electrically insulated holder and represents the cathode of a high voltage d.c. circuit and the source of evaporant material and all earthed parts form the anode. An inert gas atmosphere (usually argon) at low pressure is contained in a vacuum chamber connected to an efficient vacuum system, continuously pumping the chamber. During the process, the chamber is pumped down to a pressure in the range 10^{-3} to 10^{-4} Pa ($\sim 10^{-5}$–10^{-6} Torr). When a vacuum has been achieved the chamber is backfilled with argon to a pressure of approximately 1 to 0.1 Pa ($\sim 10^{-2}$–10^{-3} Torr). The argon influx is controlled by a variable leak valve and it can be further controlled by a selected closure of the baffle valve between the work chamber and the vacuum system. A d.c. bias of -1 to -5 kV is then introduced across the high voltage electrode (substrate) and the ground for the system. A glow discharge occurs between the electrodes which results in the substrate being bombarded by the high energy argon positive ions and energetic neutrals produced in the discharge. The argon ion bombardment effectively cleans the substrate surface by removing oxides and other surface contamination such as carbon (Kominiak and Mattox, 1977). The cleaning effectiveness can be monitored during the sputtering process by monitoring the discharge current while keeping the gas pressure and discharge voltage constant. Figure 49 shows the sputtering current density as a function of sputter cleaning time at 3 kV bias and 1 Pa argon pressure.

Since the discharge current consists partially of secondary electrons emitted from the sputtered surface, it is sensitive to the surface condition. Thus the initial high value of the current results from the presence of surface oxides since, generally, these oxides have a higher secondary electron emission coefficient than pure metals. A gradual decrease in current will occur as the oxides are sputtered from the surface until a constant minimum value is reached. At this stage, which usually occurs after 10–15 minutes, the substrate can be regarded as being free of oxides and the film deposition sequence is commenced. The coating source (filament, boat or crucible) is then energized and the coating material is vaporized into the glow discharge. The above process provides the workpiece with a uniform ion bombardment, and gas scattering along with ion deflection gives rise to the effectively high throwing power of this technique and results in a uniform coating of even rather intricate variations in surface contour. However, for a film to form, the deposition rate must exceed the

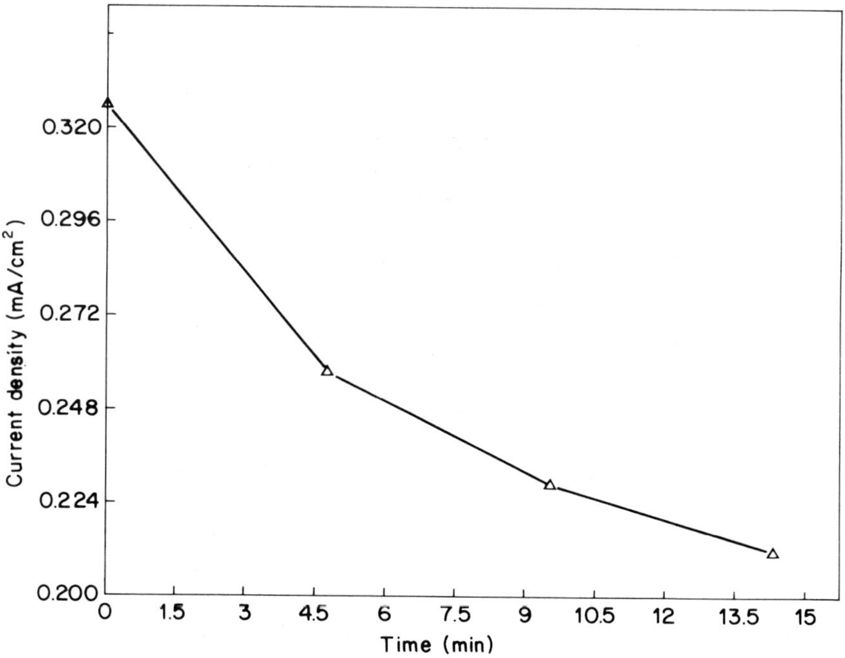

Figure 49. Relationship between the sputtering current density and sputter cleaning time at 3 kV bias and 1 Pa argon pressure

sputtering rate, and ion bombardment may or may not be continued after the interfacial region has been formed (Kominiak and Mattox, 1977). When the required thickness is achieved, the current to the vapour source and the high voltage bias are switched off and the plating process is complete.

Although the procedure described above contains all the essential features of ion plating, it has serious limitations in practical applications. For example the use of a d.c. bias as a source of a glow discharge restricts the process to conducting substrates only. For insulators the process has to include an r.f. power supply. Other limitations include the low ionization achieved with the diode ion plating. Therefore, different techniques and methods have been designed to increase the ionization efficiency. These will be discussed in detail.

4.1 ENHANCED IONIZATION SOURCES

4.1(a) Triode Ion Plating

The simplest means of supporting a glow discharge at low gas pressure is to incorporate a positive electrode to give a triode configuration (Matthews and Teer, 1981). This gives a useful increase in ionization and can lower the permissible operating pressure. Great control over ionization in triode ion plating

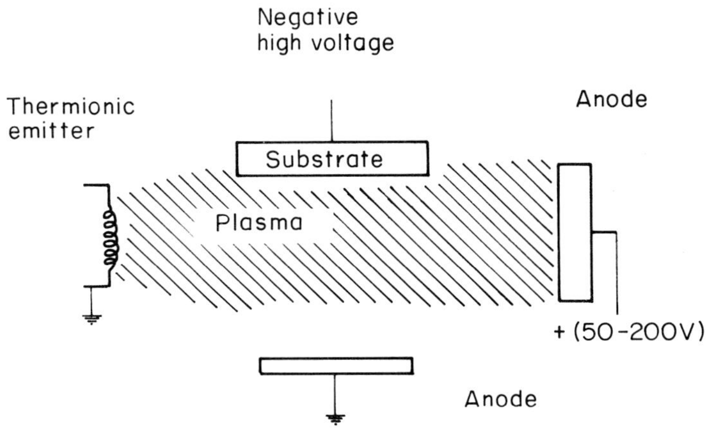

Figure 50. Thermionic assisted triode discharge

can be achieved if a separate thermionic emitter is incorporated independent of the vapour source as shown in Figure 50. In this technique the electrons maintaining the discharge are generated by thermionic emission from a hot filament and are attracted to an anode at about 100–200 V d.c. relative to the hot cathode.

The emission current can be calculated using the Richardson–Dushman equation (Chapman, 1980):

$$j = AT^2 \exp - \frac{e\phi}{kT} \tag{4.1}$$

where A is a constant equal to 120 A cm^{-2} deg $^{-2}$, T is the filament temperature, ϕ is the work function of the filament material and K is Boltzmann's constant.

The emission current is space-charge limited because of the electron space-charge surrounding the filament. Therefore, the full emission current can reach the anode if the anode bias is increased substantially or a gas at low pressure is introduced. When a gas is introduced, a glow discharge will be initiated between the cathode (filament) and the anode if the applied voltage exceeds the ionization potential of the gas. Ions produced in the discharge tend to accumulate in front of the cathode and a dark space is produced (Maissel and Glang, 1970). The dark space consists of an electron sheath in front of the cathode followed by a positive ion sheath adjacent to the plasma. This sheath neutralizes the electronic space-charge and acts as a virtual anode close to the cathode. Therefore, by applying a small positive voltage to the anode with respect to the cathode (filament), a full emission current can be drawn.

The current leaving the hot cathode is almost equal to the total current through the discharge, unlike the electron current from a cold cathode which is only about one-tenth of the discharge current (Matthews, 1980). Therefore it is possible to sustain the discharge even at very low gas pressure by increasing the voltage applied to the system. When a negatively biased substrate is introduced in the system, it is possible to control the substrate voltage or current density independently of gas pressure. This is in contrast to diode ion plating where the current density is greatly influenced by substrate bias and gas pressure. Figure 51 shows the current–voltage characteristics of diode and thermionically assisted triode discharge systems respectively. The electron path length could

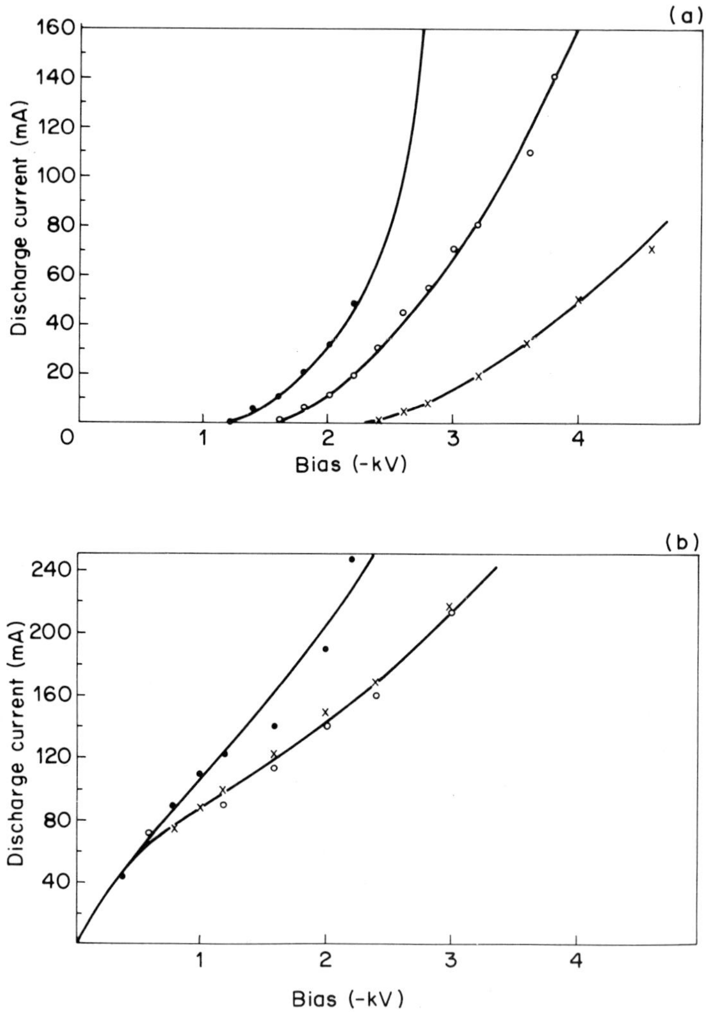

Figure 51. (a) I–V characteristics of the diode discharge
(b) I–V characteristics of the triode discharge
● 1 Pa Ar
× 0.4 Pa Ar

also be increased by the use of a small magnetic field. In this way the magnetic field acts in a way similar to the increase in gas pressure where electrons suffer more collisions with gas atoms due to the increase in their path length (Maissel and Glang, 1970).

Perhaps the most important benefit of the thermionically assisted triode system is its capability of increasing the substrate current density even at low bias levels and/or low gas pressure. This has been shown to be particularly desirable for reactive ion plating (Matthews and Teer, 1981), where the increase in ionization efficiency leads to a greater reaction. In fact, Matthews (1980) has successfully deposited dense, hard and equiaxed titanium nitride compound at substrate temperatures below 350°C by using a thermionically assisted triode discharge. It has also been reported that by using such a system, the ionization efficiency could be increased from less than 1% to over 3% and the substrate current density from 0.5 mA cm^{-2} to over 3 mA cm^{-2} (Matthews, 1980). These effects, coupled with the ability to deposit at pressures below 0.1 Pa, have been shown to produce considerable coating densification and improved reaction in ceramic coatings (Ahmed and Teer, 1982) and aluminium coating (Ahmed, 1984). Improvements in metal alloy coating properties have also been obtained by this technique (Kloos *et al.*, 1981; Minnis and Sundquist, 1981). However, when operating at gas pressures as low as 0.1 Pa, the gas scattering is reduced giving a mean free path of ~50 mm. Consequently this results in bad throwing power, and samples with complex shapes have to be rotated in order to obtain uniform coatings.

4.1(b) R.F. Coil Electrode

Another technique of increasing the ionization efficiency of a gas discharge is by employing an r.f. coil electrode in the ion plating system. Figure 52 shows an r.f. coil electrode placed between the evaporator and the substrate holder which serve as the anode and cathode respectively (Murayama, 1975; Suzuki *et al.*, 1981). The r.f. coil electrode is connected to a 13.56 MHz, 1 kW generator and the oscillating field is applied between the coil and the evaporator. This system can operate at pressures as low as 10^{-2} Pa and is used mainly for reactive ion plating. Coatings of In_2O_3, TiN and TaN on glass substrates have been achieved with this technique and found to be pinhole free, adherent and uniform. The coatings have also been claimed to have excellent electrical and physical properties (Suzuki *et al.*, 1981).

4.1(c) Other Enhanced Ionization Sources

The ionization of a gas discharge can also be enhanced by the application of a magnetic field to increase the path length of electrons as in magnetron sputtering (Mattox, 1964). Enhanced ionization can also be obtained by means of the configuration shown in Figure 53 where a three grid assembly is used between the evaporator and the substrate (Weissmantel, 1977). The hot cathode

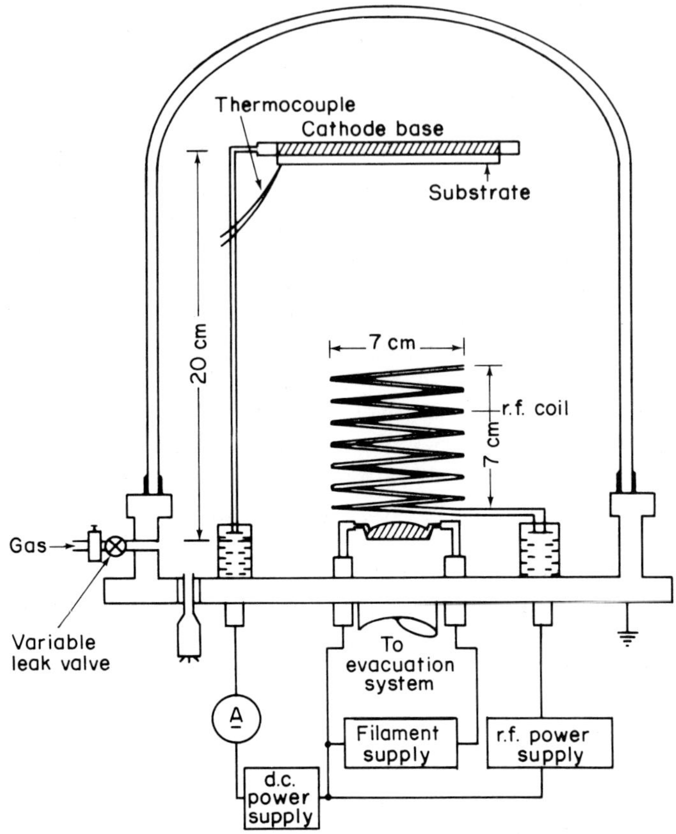

Figure 52. A r.f. ion plating system (from Murayama, 1975). (*Reproduced by permission of the American Institute of Physics*)

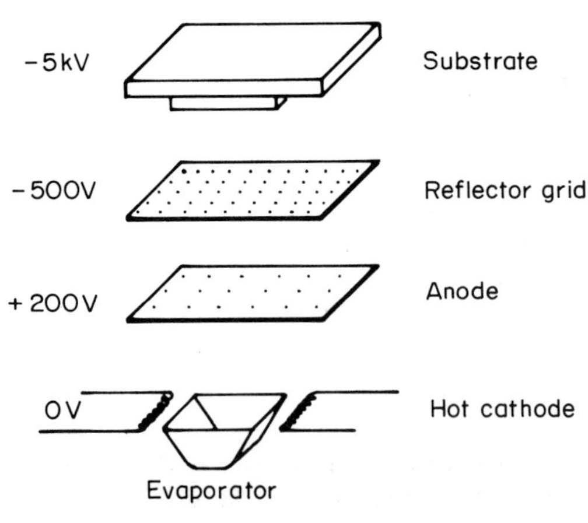

Figure 53. Schematic diagram of grid assembly to enhance ionization

and the tungsten wire reflector grid are designed so that ionizing electrons travel along pendulum orbits, thus enabling efficient ionization, and the ion beam produced is shaped electrically into bundles after passing through the grid. This method has been used by Pagel (Weissmantel, 1977) to produce hard material coatings such as chromium and titanium carbide films. The discharge can also be enhanced when using an electron beam source for evaporation if a positive electrode is used to attract secondary electrons away from the region of the crucible to enlarge the plasma region. In some cases, with easily ionized coating materials such as chromium, there may be appreciable ionization of the material being deposited so that the discharge can be sustained without the inert gas. However, the discharge in this case is not very stable and this method is generally not practical in the ion plating system. Takai (1983) has reported the utilization of a microwave discharge for high speed ion plating and reactive ion plating. A 2.4 GHz microwave discharge was operated in a simple cavity with sleeves. In this method In, InN and In_2O_3 were prepared at low temperatures.

4.2 PROCESS CONTROL

In order to attain reproducible high quality coatings, many parameters have to be adjusted and controlled for different ion plating units. Once these have been defined the process can be used repeatedly to give reproducible coatings. Some of the parameters are:

(a) Substrate material.
(b) Substrate cleaning.
(c) Substrate jigging and geometry.
(d) Substrate temperature.
(e) Pressure measuring technique.
(f) Deposition rate and thickness monitor.
(g) Exposure to atmosphere during loading and unloading cycle.
(h) Substrate voltage and current.
(i) Initial system vacuum.
(j) Purity of sputtering gas.
(k) Sputtering gas pressure and flow rate.
(l) Sputter cleaning time.
(m) Substrate–source distance.

Some of these parameters can be monitored during the ion plating process to ensure better reproducibility. Others have to be adjusted prior to loading into the vacuum chamber. These parameters will be discussed in detail.

4.2(a) Substrate Material

The surface finish and nature of substrate are very important factors which may affect the properties of coatings deposited on them. Some materials may alloy with the coating while others might be sensitive to temperature, bias or ion bombardment. Certain types of plastics may degrade by the ultraviolet light emitted in the plasma or as a result of ion bombardment. In this case care should be taken to protect the plastic substrate from UV light or severe ion bombardment. Surface finish may also affect the appearance of the final finish as well as the adhesion and structure. Substrate surface irregularities such as inclusions or debris particles that fall on the substrate or growing coating can cause preferential nucleation and runaway growth (Patten, 1978), as shown in Figure 54. Substrate surface roughness may cause shadowing and result in a non-uniform open structure with bad mechanical properties (Bunshah, 1982). Such shadowing occurs because high points on the growing surface receive more coating flux than valleys do, particularly when a significant oblique component is present (Thornton, 1974; 1979).

4.2(b) Substrate Cleaning

Cleaning is generally defined as the process of removing undesired matter from the surfaces of the parts prior to coating. It usually involves the breaking of the bonds between contaminant particles and the substrate surface by utilizing mechanical or chemical methods (Mattox, 1978). Although substrate cleaning is usually carried out in the glow discharge cycle of the ion plating process, it is important to have an initially clean surface to minimize system contamination, reduce cleaning time and obtain a highly adherent coating with good structure (Glang *et al.*, 1970). However, the selection of a suitable cleaning process depends mainly upon the nature of the substrate being cleaned, the soil to be removed and the size and number of parts to be cleaned. While light polishing, scrubbing and ultrasonic cleaning with detergent or solvent may be adequate in the laboratory, harsh treatment with grit blasting and chemical etching will be

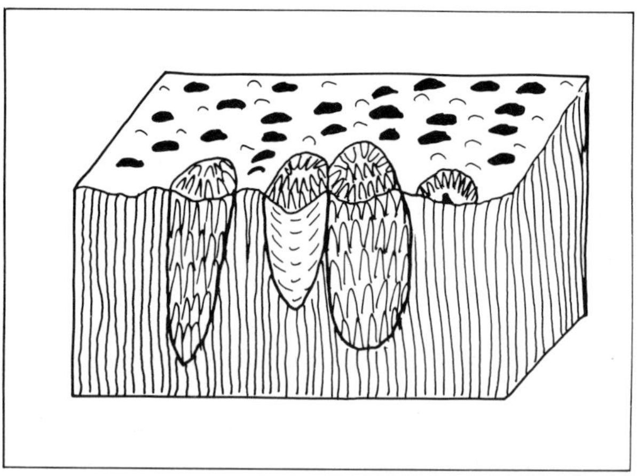

Figure 54. Effect of surface finish and inclusions on nucleation and runaway growth

required in industrial production. Typical soils existing on parts to be coated such as fasteners and cutting tools are: drawing and stamping lubricants, cutting fluids, heat treatments, scale and oxides and fingerprints. In the case of non-metallic parts such as glass and ceramics, the contaminants may be mainly water containing oxidic polishing residues, inorganic and organic dust particles or films consisting of oil and grease or a combination of these. Therefore, the removal of such soils requires more than one cleaning operation to achieve a certain degree of cleanliness necessary for good coatings. The cleaning operations are usually conducted outside the vacuum chamber while the final cleaning process in the glow discharge is usually carried out inside, prior to the evaporation of the required coating material.

4.2(b.1) Cleaning Metals

Cleaning of metal parts depends on many factors including nature of metal, type of soils and the size of the parts to be ion plated. While ferrous metals, stainless steel and magnesium alloys are not affected by alkaline solutions, aluminium alloys, zinc and cadmium parts may be attacked or tarnished by such solutions. Copper is merely stained by alkalis, yet severely attacked by oxidizing acids such as nitric acid. Titanium and zirconium have excellent chemical resistance to alkali and acids but are attacked severely by acid fluorides. Therefore, a properly selected cleaner must be used for each metal. Removing soils from metal parts depends on the type of soil and it could be eliminated by using a combination of mechanical and chemical methods. Mechanical methods involve the use of abrasive cleaning with grit blasting, while chemical methods involve the use of detergent, solvents, and alkaline or acid cleaners. Table 3 shows typical cleaning processes used for different metals prior to ion plating with different coatings. These methods are adopted in the laboratory or on an industrial scale.

4.2(b.1.1) Dry Blasting

Dry blasting is used extensively as part of a full cleaning process to remove scale and rust from grease or oil-free metal parts. For this reason, dry blasting is usually carried out after solvent vapour degreasing and drying. The most versatile dry blasting machine used in industry is the air blast, where the blasting medium is propelled by a stream of pressurized dry air. Aluminium oxide is the most widely used medium for this process, but silicon carbide, glass beads and other materials are occasionally used for special applications (Wick and Velleux, 1985).

Good results can be obtained with media particles of around 200 μm at an air pressure of 30–60 p.s.i. Finer powders are usually used for cleaning and polishing. It is important not to re-use the powder since it will lower the cutting action and the contaminations may clog the orifices. Pressurized air used should be dry and filtered. After grit blasting, the parts should be flushed well with dry nitrogen gas or oil-free filtered air to remove excess grit, dust and

Table 3.

Ref. No.	Substrate Material	Type of Coating	Cleaning Process
(Wick and Velleux, 1985; McDonnell Douglas, 1985)	Steel	Al, Cr	Vapour degrease → alkaline rust removal → grit blast → isopropyl alcohol → coat
(McDonnell Douglas, 1985)	Titanium/titanium alloys	Al	Vapour degrease → chemical degrease → grit blast (or pickling) → solvent wash → dry → coat
"	Al	Al	Vapour degrease → grit blast (or chemical degrease) → pickling → rinse → dry → coat
(Helmersson *et al.*, 1985)	High speed steel	TiN	Polish → ultrasonic in trichloroethylene → acetone → ethanol → coat
(Holland, 1956; Ahmed, 1986)	Glass, silica	Au, Ag, Al	Ultrasonic clean in detergent → rinse in deionized water → rinse in isopropyl alcohol → dry in nitrogen gas → coat

contaminants prior to coating. Any type of contaminants that stay adhered to the blasted surface may produce surface defects in the coating (McDonnell Douglas, 1985).

4.2(b.1.2) Chemical Cleaning

Chemical cleaning is usually used with or without dry blasting to provide a suitable surface for subsequent finishing by ion plating. Chemical cleaning depends upon the chemical action between the cleaning material and the contaminants or substrate (Wick and Velleux, 1985). Different types of chemicals such as solvents, solvent emulsions, alkaline and acid cleaners are often used in the laboratory or in industrial production. However, there is always a danger of leaving residues on the surface if solvents are used and hydrogen embrittlement if acid cleaners are used due to the absorption of hydrogen during the cleaning process (McDonnell Douglas, 1985). The most general solvent cleaners used in vapour degreasing to remove oil and grease are trichloroethane or perchlorethylene. Alkaline rust removers are generally used to remove rust from steel, aluminium, zinc and tin prior to ion plating. The alkaline solutions may contain occasionally surfactants to aid removing light soils or to soak cleaner films. Acid cleaners are designed to remove oxides, rust and corrosion products. Nitric acid alone or mixed with hydrochloric acid is used to deoxidize and remove scale. However, care should be taken to rinse the parts well in warm deionized water to remove extra acid followed by good drying in a centrifuge to remove water. Warm air at a maximum temperature of 70°C can also be used instead of the centrifuge to dry the parts. Chemically cleaned parts are very active and should be loaded into the vacuum chamber soon after cleaning to prevent rusting or contamination. It should be mentioned at this point that on certain parts such as steel, chemical cleaning cannot be a good substitute for grit blasting due to residual contaminants and etching that will effect adhesion. Figure 55 shows a comparison between aluminium ion plated steel fasteners cleaned chemically with 10% hydrochloric acid and fasteners cleaned with grit blasting prior to coating. The adhesion and the final finish of the grit blasted fasteners looks better than those cleaned with acid.

4.2(b.1.3) Ultrasonic Cleaning

Ultrasonic cleaning can be used in conjunction with different types of cleaners to remove highly adherent contaminations. The other advantages of the technique are the cleaning of small precision parts of complex configurations and parts of sealed units. The most commonly used cleaners are chlorinated hydrocarbon solvents, detergent and water. In ultrasonic cleaning, shock waves generated in the solution enhance the dissolution of contaminants and carry them away from the part surface. Mechanical vibrations induced in the parts aid in loosening contaminants such as particulate matter flakes (Mattox, 1978). However, because ultrasonic cleaning is more expensive than other methods, it is usually used when other methods prove inadequate.

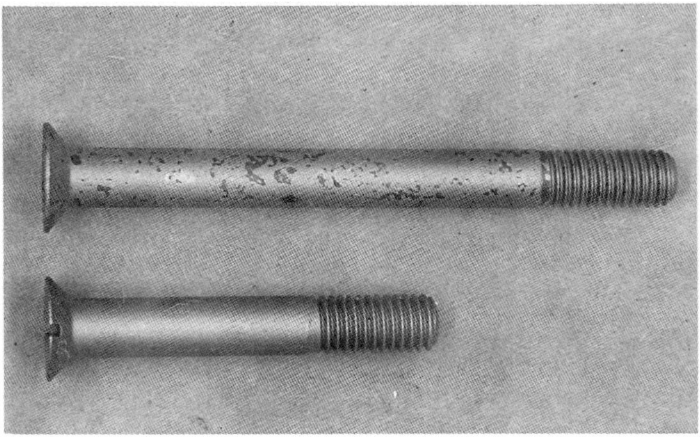

Figure 55. Comparison between aluminium ion plated steel fasteners cleaned chemically and fasteners cleaned with grit blast prior to coating

4.2(b.2) Cleaning Non-Metals

Cleaning non-metal parts such as glass, plastic or ceramics requires a different approach to that for metals due to their mechanical and thermal instability. In general, glass surfaces can be cleaned successfully by scrubbing with a cotton wool cloth dipped in a detergent water solution, followed by a good rinsing in distilled water and drying with a tissue paper (Pulker, 1984). If the surface is rough or porous as in sintered ceramics, the cleaning should be assisted by ultrasonic agitation in a solvent or detergent followed by a good rinsing in distilled water. The selection of a suitable solvent depends mainly on the type of contamination but in general solvents such as alcohols, ketones and chlorinated and fluorinated hydrocarbons are used to clean glass or semiconductor lenses. However, when solvent is used, the surface should not be allowed to dry before the final rinse with distilled water. In the optical industry, glass and germanium lenses are usually cleaned with cotton wool dipped in a mixture of carbon black or chalk and alcohol (Macleod, 1969). The chalk should be removed from the surface after this treatment with pure water or alcohol. Vapour degreasing is usually used as a last stage in the cleaning process of non-metal parts to remove grease and oil from the surface prior to loading into the vacuum chamber. For further details on the cleaning process, the reader is referred to the books by Pulker (1984) and Holland (1956).

4.2(b.3) Glow Discharge Cleaning

In the ion plating process the parts to be coated are usually given a final sputter cleaning treatment with the glow discharge inside the vacuum chamber prior to coating. Discharge sputter cleaning is generally performed either in argon or oxygen atmosphere to remove hydrocarbons and other contaminants from the surfaces. Cleaning is carried out by the direct heating effect on the substrate due to ionic bombardment. It also involves other mechanisms such as disorption of impurities from the surface by electrons, neutrals or ion bombardment, or volatization of organic residues by chemical reaction with dissociated oxygen (Pulker, 1984). When oxygen is used for sputter cleaning, care must be taken not to oxidize the substrate or the evaporated material. For cleaning insulated parts such as glass or plastic, the parts can be biased with an r.f. power or can be separated from the glow discharge circuit. A negatively biased wire mesh with large perforated holes may be used adjacent to the insulated surface so that the energetic particles pass through it and bombard the surface. The mesh should be coated with the same coating material to prevent contaminating the surface. Furthermore, plastic parts should not be sputtered for too long otherwise deterioration may occur due to heating by secondary electrons. Beside its cleaning effects, glow discharge also assists the modification of the surface by inducing surface disorder or by causing chemical activation and cross-linking in plastics. The latter increases the surface strength of the polymer and reduces the amount of undesirable low molecular weight components (Mattox, 1982b).

4.2(b.4) Storage

Clean parts are usually very active and may react quickly with the contaminated atmosphere. Therefore, it is always recommended to coat the parts within a short period of cleaning. If the parts are to be stored for a certain time prior to coating, then it is necessary to identify the undesirable contaminants and to eliminate them from the storage environment. The most common contaminants are hydrocarbons, dust, chemical vapours and water vapour (Pulker, 1984; Mattox, 1982b; Holland, 1956). Contact with dust may be minimized by storing the parts in a closed container or dessicator, or in a clean bench. Contamination with hydrocarbon vapours can be reduced by storing the parts in freshly oxidized aluminium containers which preferentially absorb the hydrocarbons (Mattox, 1982b). A disadvantage of this technique is the extra steps involved in periodically stripping and re-oxidizing the metal surface of the container. If rust spots appear on the cleaned surfaces of ferrous alloys during storing, then the parts should be re-cleaned before coating (McDonnell Douglas, 1985). However, as a general rule, the cleaned parts should be coated immediately or within 30 minutes after cleaning and should be loaded carefully in the coating chamber using plastic gloves.

4.2(c) Substrate Jigging and Geometry

The way the substrate jigs in the vacuum chamber varies from simple stationary holders to complex rotating barrels or carousels depending, in general, on the substrate geometry and number of items to be coated. The main aim is to coat

Figure 56. Rotating carousel to ion plate missile cases with aluminium (*Reproduced by permission of McDonnell Douglas Corp.*)

the required number of items as economically and uniformly as possible. Thus, for coating large numbers of small parts, rotating barrels can be used while rotating jigs can be employed to coat large cylindrically shaped parts such as missile cases (Figure 56). In such parts the jigs are rotated 180° to allow complete coverage of the parts with the coating material without interrupting the coating cycle (Muehlberger and Reilly, 1985). Drills and cutting tools can be ion plated with titanium nitride using a planetary rotating carousel (Figure 57). However, because ion plating is not confined to line-of-sight deposition, small parts with complex shapes can be coated with a reasonably uniform coating without the need for rotation. It should be emphasized at this point that rotating jigs are usually biased to a high negative voltage and care should be taken to select suitable insulators to isolate the jig from the earthed surroundings. Insulators may fail and become short-circuited as a result of wear due to rotation or high torques or as a result of being coated if they are not well masked. When a rotating jig is used to hold certain items, the coating time for a particular thickness depends mainly on the evaporation rate, the rotation speed and surface area of the item to be coated as well as the gas pressure. In order to ion plate large areas of plastic sheets, it is necessary to achieve high deposition rates without increasing the substrate temperatures (Ridge *et al.*, 1981). This can be done by pulling the sheet substrate tightly over a very smooth cooled drum and

Figure 57. Rotating carousel to ion plate drills with titanium nitride coating. (*Reproduced by permission of TECVAC Ltd, UK*)

transferring the plastic from an initial roll to a final roll over the rotating drum (Ridge *et al.*, 1981). The drum has to be isolated electrically from the earthed surroundings.

4.2(d) Substrate Temperature

The increase in substrate temperature during the ion plating process can be related to the influence of four main sources of heat input. The first is ion bombardment of the substrate which may result in energy input as high as 100 eV per atom (Pulker, 1984; Lamont and Lang, 1970; Lau *et al.*, 1972). Radiation emitted from the glow discharge including energetic electrons, ultraviolet and infrared radiation may contribute energies in the range 2 to 10 eV per atom depending on the discharge intensity (Lamont and Lang, 1970; Lau *et al.*, 1972). Heat radiation emitted from the evaporation source depends mainly on the source temperature and the emissivity of the molten source and can be as high as 1000 eV per atom for a molten tungsten source operating at 3800 K (Matthews, 1980). Heats of condensation of coating atoms or molecules on the substrate may contribute input energy in the range 3 to 9 eV per deposited atom. Therefore, these main sources of heat input will result in a high substrate temperature if the power dissipation is poor. This can be reduced by attaching the sample to a water-cooled holder, by stopping the coating cycle and flushing the vacuum chamber with nitrogen or inert gas, by shortening the sputter cleaning cycle or by changing the ion energy spectrum to improve the sputtering efficiency (Lamont and Lang, 1970). In certain applications it is necessary to heat up the sample in order to improve the adhesion and structure or to impart greater stress in the coating through differential contraction on cooling (Mattox, 1982b). This is usually accomplished by radiant heating from quartz lamps, or by direct conduction from a resistance heated holder.

4.2(e) Pressure Measurement Technique

The monitoring of the inert gas environment is of considerable importance particularly in reactive ion plating where the partial pressure of the reactive gas must be related to the deposition rate to maintain the desired degree of stoichiometry. In general, the pressure in the backing line, the roughing line and the vacuum chamber is monitored by Pirani or thermocouple gauges in the backing and roughing lines and an ionization gauge in the chamber (Goode and Burden, 1979). The Pirani gauge operates in the region 0.1–5 Pa by measuring the change in resistance of the heated filament, which forms one of the arms of a Wheatstone bridge, within the vacuum system (Harris, 1983). The resistance of the filament depends on its temperature, which in turn depends on the heat transfer. The Pirani gauge, however, has two fundamental weaknesses when used in the ion plating system. The first is that most Pirani gauges are calibrated for dry nitrogen, not argon, and are accurate under normal vacuum conditions where small quantities of other gases and water vapour are present. The second weakness is that its sensitivity and hence accuracy and reproducibility are poor in the 2.5 Pa region. To overcome these problems, the Pirani gauge should be calibrated for argon gas (or any other gas used in the system) and an argon-calibrated high pressure ionization gauge can be used in the vacuum chamber to measure the argon pressure during the plating process. A sensitive capacitance manometer gauge has also been found to give reliable pressure measurements and can provide a control signal to the gas admittance system (Figure 58). However, they are very sensitive to vibration. In the present of r.f. discharge, ion gauge and capacitance manometers give misleading results since the r.f. signal stimulates excess ionization in the ion gauge head. Therefore, these gauges should be well shielded electrically to give reliable results. To measure the partial pressures of contaminant gases or reactive gases during

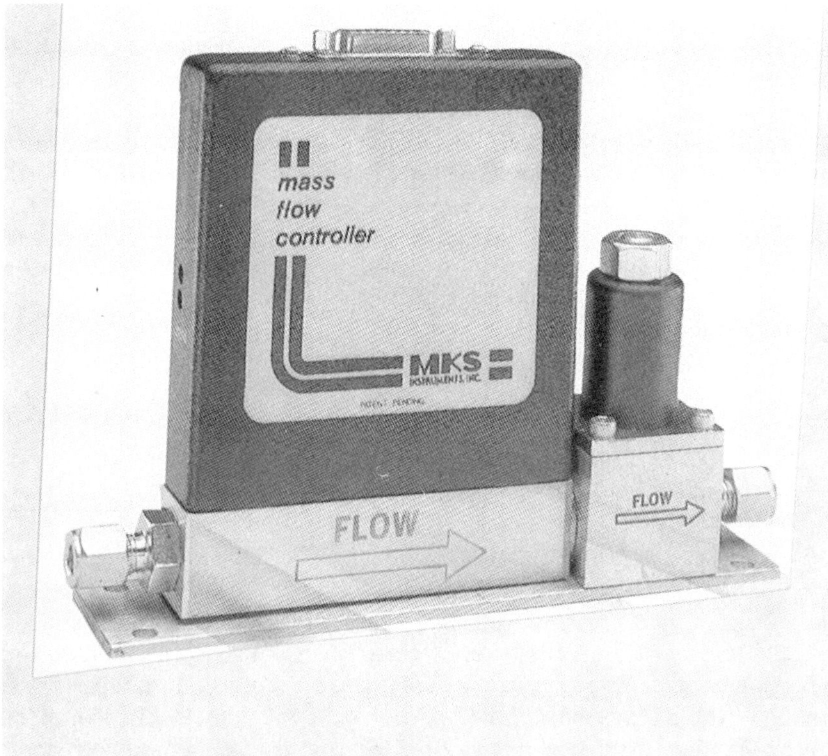

Figure 58. MKS gas flow controller. (*Reproduced by permission of Chell Instruments Ltd, UK*)

reactive ion plating it is useful to use a differentially pumped mass spectrometer in the vacuum chamber (Mattox, 1965).

4.2(f) Deposition Rate and Thickness Monitor

The deposition rate and thickness can be measured using several techniques. The knowledge of the actual deposition rate and thickness during the ion plating process would enable corrective action to be taken by including minor variations in the process variables. One of the methods used to measure the deposition rate is by monitoring the weight gain on a surface placed near the substrate by means of a microbalance or quartz crystal oscillator. The crystal oscillator monitor utilizes the piezoelectric properties of quartz. The resonance frequency induced by an a.c. field is inversely proportional to crystal thickness and therefore to the mass of the deposited film (Stockbridge, 1966). The change in the frequency of the crystal exposed to the vapour stream is compared to that of a reference crystal. An example of crystal monitor is shown in Figure 59. However, when using a crystal thickness monitor in an ion plating system,

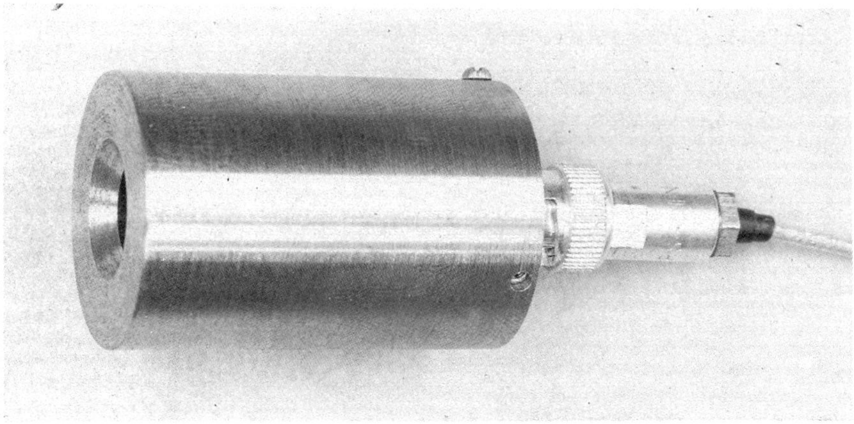

Fig. 59 Crystal thickness monitor

some problems arise such as excessive heating by charged energetic particle bombardment and the thick deposition which is typical of the ion plating system. Heating by ion bombardment causes frequency instabilities. A grounded front electrode of the sensing crystal prevents electron bombardment, and the plasma can be kept away from the crystal by using a grounded aluminium grid in front of the aperture of the quartz holder (Chabicovsky *et al.*, 1970), although this may reduce the crystal sensitivity. Another convenient method to reduce the ion bombardment effect on the crystal is by employing an intense local magnetic field of about 1000 gauss in front of the crystal holder aperture to deflect the charged particles (Lu, 1978). For thick coatings, the use of a crystal oscillator is not very practical since it has to be cleaned or replaced each run. To overcome this problem a grid could be used in front of the holder aperture to reduce the amount of evaporant reaching the crystal surface, or the holder could be made in the shape of a long tube to increase the distance between the holder aperture and the crystal surface. Other physical properties such as reflectance, transmittance or electrical resistance can be used on a dummy substrate to monitor the film thickness (Pulker, 1984; Holland, 1956).

By using a suitable monitor, it is possible to feed back the signal to control the power input to the electron beam gun, or resistance heated boats power supply and, therefore, to control the rate of deposition. This is important in the case of multi-layered optical coatings (McLeod, 1969). If an r.f. system is used, then the detector leads and all electronic circuits must be well shielded electrically.

4.2(g) Exposure to Atmosphere During Loading and Unloading Cycle

It is important not to leave the vacuum chamber exposed to the atmosphere for long periods during loading and unloading of samples. Water vapour and other contaminants usually adsorb on the metal surface to a thickness of several monolayers. During the pump down the gas that is present in the chamber and the gas which is physisorbed on the walls are removed very quickly (Glang *et al.*, 1970). Thereafter the pressure is determined by a balance between the net rate of deposition and diffusion of gas from the substrates and the walls of the chamber and the pumping speed. Strongly chemisorbed species release at such slow rates that they do not contribute significantly to the pressure. However, the water vapour, which is bonded with intermediate energies (~ 1 eV), releases slowly and dominates the chamber pressure during typical pump down times (10 hr) in room temperature chambers (Glang *et al.*, 1970). Therefore, the outgassing rate from the chamber walls and substrates depends to the first order, only on the time that the chamber is pumped while the pressure depends on the pumping speed. Furthermore, the dissociation of water vapour during the glow discharge produces hydrogen and oxygen. The presence of oxygen will lead to the formation of metal oxide which will subsequently cause the formation of metal films of rough grain structure and higher electrical resistivity (Glasser and Szigeti, 1983). The presence of hydrogen, on the other hand, may affect the deposition rate. Significant variations in deposition rate from run to run and within a run have been attributed to these effects (Glasser and Szigeti, 1983; Stern and Coswell, 1967). In certain applications it is recommended to run the ion plating process in an air-conditioned room to keep down the humidity level and to prevent dust from entering the vacuum chamber during loading and unloading (McDonnell Douglas, 1985).

4.2(h) Substrate Voltage and Current

During the sputter cleaning cycle, the substrate is biased negatively relative to the plasma potential. Contamination is thereby removed from the substrate surface by sputtering. Typical sputter cleaning conditions in a diode ion plating unit are: current density of 0.5 mA/cm^2; voltage 1–2 kV and sputtering time of 5–15 minutes depending on the type of substrate. An r.f. bias is required if the substrates are non-conducting. It is important to ensure a clean surface prior to coating. During the coating cycle, it is possible to lower the bias to make the deposition rate higher than the sputtering rate. These conditions must be kept the same, from run to run, to obtain a reproducible coating.

4.2(i) Initial System Vacuum

The pumping system must be capable of rapid pump down to the required vacuum, usually 10^{-3}–10^{-4} Pa, and be capable of maintaining the vacuum during deposition. Outgassing from the chamber walls, that are often covered with thick powdery deposits having very large surface areas, can be very great, and is increased by the radiant heat from the vapour source (Teer, 1983). Empirically determined monographs are available, which relate the pump sizes and pump down times for various vacuum systems and materials. It is highly recommended that the chamber be heated to about 50°C after it has been open to the atmosphere to prevent condensation of water vapour, leading to excessively long pump down cycles.

4.2(j) Purity of Sputtering Gas

Gases used in the ion plating process are usually inert gases or mixtures normally provided from a commercial source, whose purity is known. In general, the highest purity gas that is economically feasible (~99.99%) should be used. Impurities of more than 0.002% in the used gas can cause undesirable chemical reaction with the evaporated material and may cause an appreciable reduction in deposition rate (Mattox, 1965). Gases such as carbon dioxide and water vapour are decomposed in the glow discharge to produce oxygen. Carbon monoxide shows a very slight tendency to increase the deposition rate, probably because it removes traces of residual oxygen from the gas (Maissel and Glang, 1970). Nitrogen has essentially no effect. Inert gases may be purified by flowing them over a hot active metal (e.g. Ti, Zr, etc.) chip bed or by the use of a getter deposition system where the incoming gases are exposed to a freshly deposited film before they enter the discharge region (Mattox, 1982a).

4.2(k) Sputtering Gas Pressure and Flow Rate

The control of the sputtering gas pressure and flow rate enables accurate control over coating thickness uniformity through the effect of throwing power and the control of chemical reaction to give the desired composition. Inert or reactive gases can be accurately fed to the vacuum chamber by using electronic mass-flow sensors controlled by feedback from a pressure gauge such as capacitance manometer and connected to servo-actuated solenoid gas valves (Hmiel, 1985). The capacitance manometer can provide total absolute pressure measurement independent of gas composition and a high degree of stability. If the pressure deviates from the fixed setting, the pressure gauge sends a signal to the flow meter to alter the gas flow rate as appropriate.

4.2(1) Sputter Cleaning Time

Sputter cleaning by the glow discharge is considered to be the final process to atomically clean the substrate prior to coating. Sputter cleaning time depends mainly on the degree of cleanliness of the substrate, the discharge gas pressure, the current density and substrate bias. During sputter cleaning the sputtered contaminants should be flushed away by using a high system gas throughput. Care should be taken not to over-sputter materials such as plastic to protect it from degradation. In general, the discharge current can be used to indicate cleaning condition. The current usually drops to a stable value as contaminants are removed from the discharge (Mattox, 1973).

4.2(m) Substrate-to-Source Distance

In order to ion plate coatings of equal thickness on the front and back surfaces of a stationary substrate, either the distance between the source and the substrate or the gas pressure can be varied. In certain applications where a cold hollow cathode gun is used to evaporate the coating material, only the substrate-to-source distance is varied (Wan *et al.*, 1971; Dugdale, 1966). Work carried out by Wan *et al.* (1971; 1973) showed that for ion plating in argon, the ratio of the coating thickness on back and front surfaces of a flat plate increases with the increase in substrate-to-source distance (Figure 60). Figure 61 shows the effect of substrate-to-source distance on the microstructure and thickness of ion plated Al coating at 2 kV bias and 1 Pa argon pressure. In most applications, however, it has been found that a substrate-to-source distance of 15–20 cm is ideal to eliminate heating effects, spitting effects, where evaporant droplets jump from the heated boats and stick to the surface of the coating (Figure 62), and to improve the adhesion of the coating to the substrate (Chambers and Carmichael, 1971; Abu Zeid, 1982; Patten, 1978).

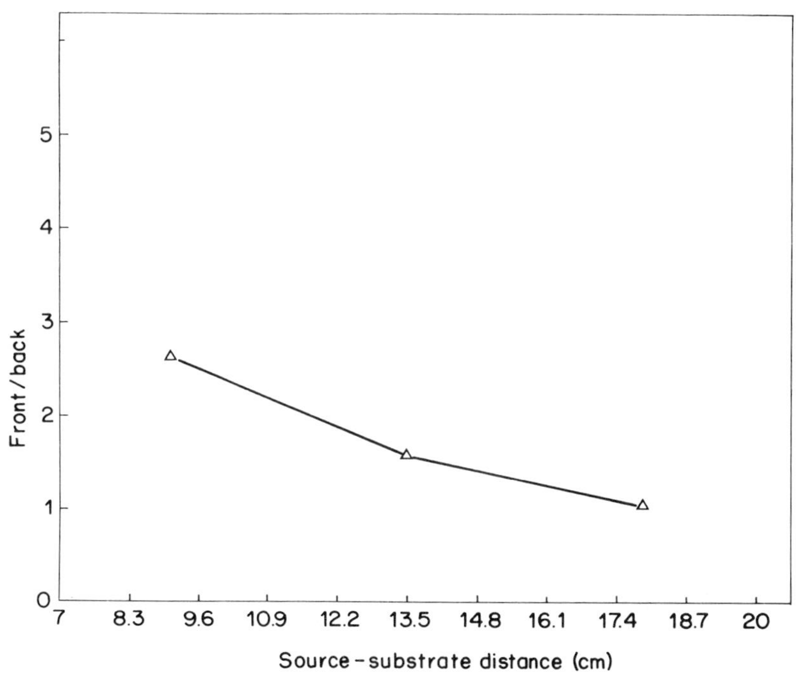

Figure 60. Effect of source-to-substrate distance on coating thickness on back and front surfaces of a flat plate at 2 kV substrate bias and 1 Pa argon pressure

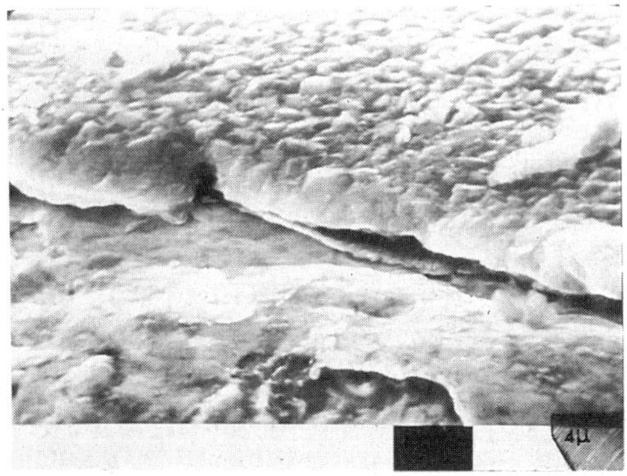

23 cm; 1 Pa; 2 kV

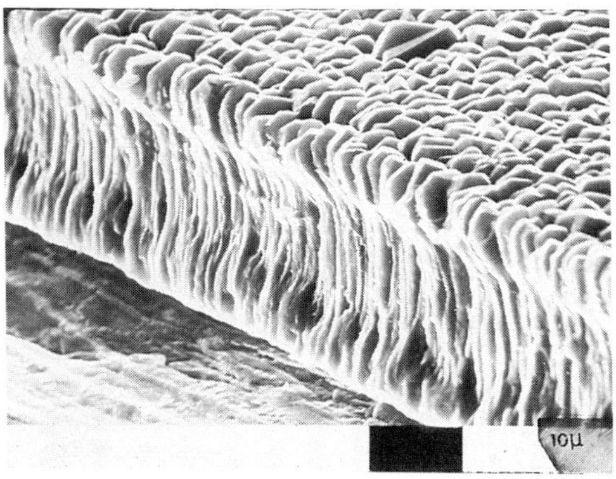

10 cm; 1 Pa; 2 kV

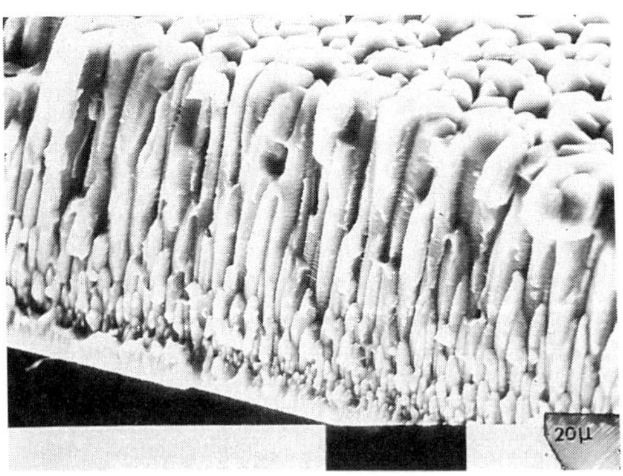

5 cm; 1 Pa; 2 kV

Figure 61. Effect of source-to-substrate distance on the microstructure of ion plated aluminium coating at 2 kV bias and 1 Pa argon pressure

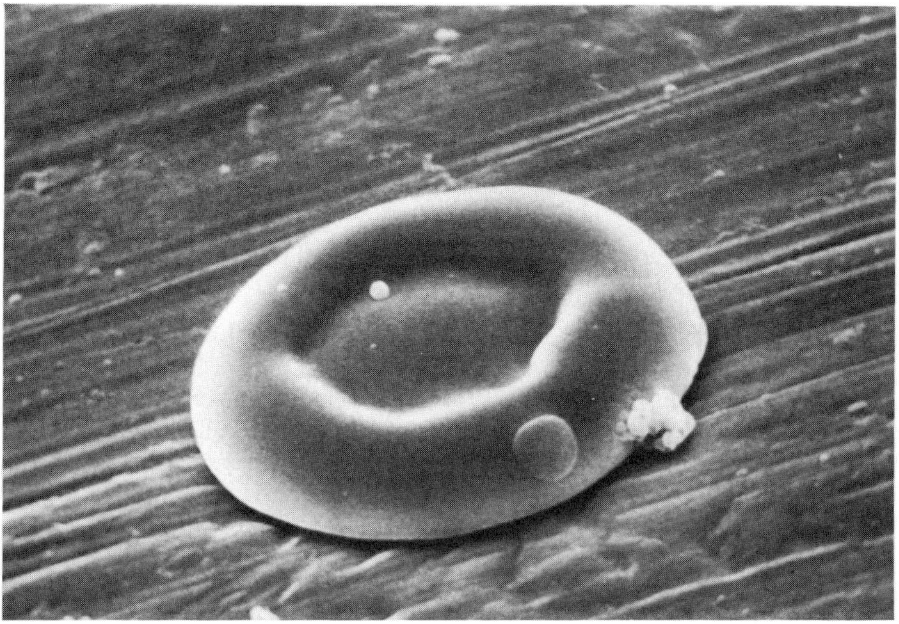

Figure 62. Spitting effect

Chapter 5
Reactive Ion Plating

5.1 GENERAL CONSIDERATIONS

Reactive ion plating as shown schematically in Figure 63 is ion plating in the presence of a reactive gas. The evaporant atoms can still be injected into the glow discharge by resistance heating, by using electron beam gun, or arc evaporation source (Figure 64). For reactive ion plating it is not necessary to use an atmosphere of pure reactive gas but a mixture of inert gas and a small quantity of reactive gas can be sufficient and can often be advantageous (Auwarter, 1960). When the reactant gas is introduced into the plasma during deposition in controlled amounts, it undergoes dissociation and ionization of the atoms of the reactant gas molecules. The reactions which are of importance in reactive ion plating process electron impact ionization, excitation and dissociation (Bunshah and Deshpandey, 1985). In this way the less reactive gases can be reactivated by the glow discharge so that nitrides, carbides, borides and other mixed compound films can be produced (Holland, 1977; 1978). For example, if titanium is evaporated into a discharge formed in a mixture of argon and nitrogen, then titanium nitride coatings can be deposited under certain conditions according to the reaction.

$$2\ \mathrm{Ti} + \mathrm{N_2} \rightarrow 2\ \mathrm{TiN} \qquad (5.1)$$

Similarly, if titanium is evaporated into a discharge formed in a mixture of argon and oxygen, titanium dioxide can be produced according to the reaction.

$$\mathrm{Ti} + 0_2 \rightarrow \mathrm{TiO_2} \qquad (5.2)$$

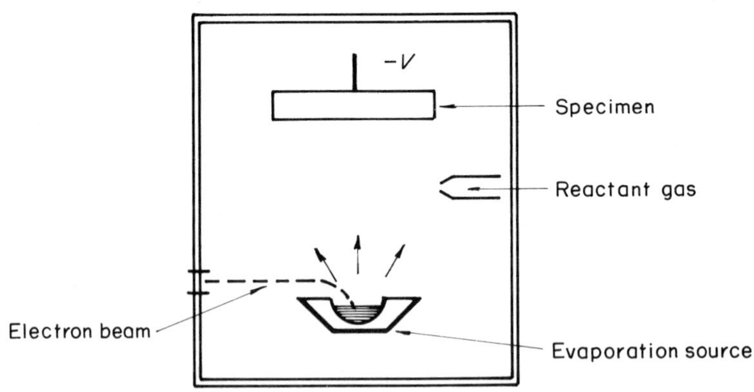

Figure 63. Schematic diagram of reactive ion plating set-up

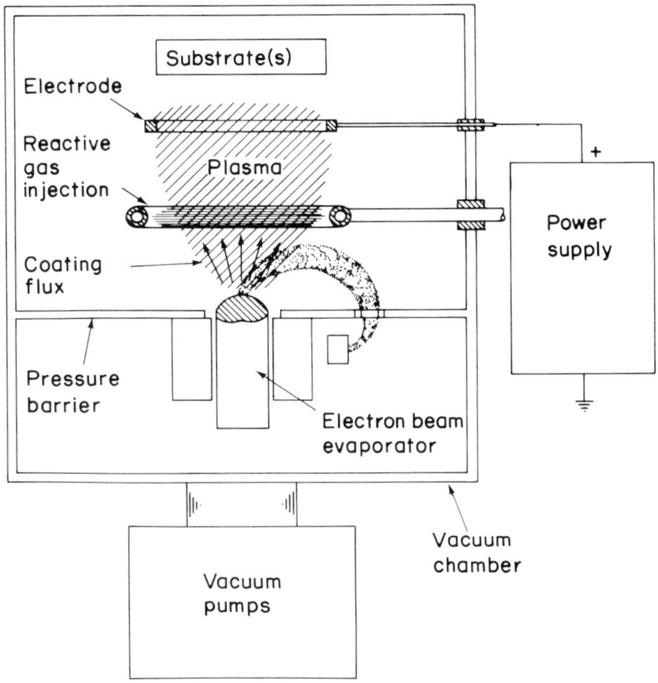

Figure 64. Activated reactive evaporation (ARE) using an EB evaporation source (from Bunshah, 1981). (*Reproduced by permission of Elsevier Sequoia SA*)

Reactive ion plating can be considered as a straightforward process if only one compound between the evaporant atoms and the reactive gas exists. However, in most cases a series of compounds may be formed depending on the conditions prevailing during the plating process. It has been found that the particular phase formed in the reaction depends on many complex important factors such as the types and concentrations of metastable species, ionized species, energetic neutrals and the free energy of formation of the compound (Bunshah and Deshpandey, 1985). Obviously, during the reactive ion plating process the analysis and control of such factors becomes even more complex. The degree of ionization of reactive gas seems to play an important role in achieving the required stoichiometry. It has been reported by Matthews and Teer (1981) that at fixed evaporation rate and reactive gas flow rate, no reaction takes place when diode ion plating is used. However, as the degree of ionization increases from 0.1% to 0.3%, by employing a thermionically-assisted triode ion plating, a cohesive and stoichiometric TiN compound is formed. Similarly, the author succeeded in forming near-stoichiometric Al_2O_3 by enhancing the discharge in the ion plating process. Hollow cathode guns (Sato *et al.*, 1978), induction heated sources (White, 1973), magnetron sputter sources (Ridge *et al.*, 1982) and triode assisted discharges (Ahmed and Teer, 1982; Matthews and Teer, 1979) have all been used to enhance ionization and to reactively ion plate compounds. Besides increasing the probability of chemical reaction, the enhanced discharge and the resultant intense ion current density to the substrate improve the compound coating structures so that dense, equi-axed structures are formed at temperatures much lower than those required for chemical vapour deposition (CVD) (Perry and Archer, 1980).

During the reactive ion plating process, the chemical reaction between the reactive gas species and the evaporant atoms may take place at one or more of three possible sites in the discharge chamber. The reaction may take place in the vapour phase or the substrate or coating surface, or on the evaporant or target surface (in the case of sputter ion plating). Gas phase reaction seems to be unlikely because of problems of conserving momentum while also dissipating the heat of reaction (Pulker, 1984). The bulk of evidence, however, suggests that reaction occurring at the substrate surface is the most important mechanism in the formation of a compound coating (Vossen and Kern, 1978). On the other hand, reactions occurring at the evaporant source seem not to pose a real problem in reactive ion plating since any compound formed on the surface of the evaporating metal is likely to be dissociated as a result of the high source temperature, particularly under the impact spot of the electron beam or cathode arc (Bunshah and Deshpandey, 1985). Furthermore, the stirring action in the liquid metal pool pushes away any compound layer, thus providing a metallic surface at the top of the liquid pool (Bunshah and Deshpandey, 1985).

However, for low melting metals, care should be taken to avoid the reaction between the reactant gas and vapour source surface since this may contaminate the melted material and lower the evaporation rate (Pulker, 1976). In reactive sputter ion plating, the target/gas plasma interaction leads to the formation of a compound layer on the target. This effect is generally known as target poisoning and it plays an important role in the growth of a compound coating. The poisoning effect can reduce the deposition rate due to the low sputtering yield of the compound formed on the target surface (Heller, 1970; Maniv and Westwood, 1980). Other effects also occur during the introduction of the reactive gas such as the abrupt drop of the deposition rate as gas flow rate increases (Figure 65). The critical flow rate at which the abrupt transition occurs indicates a transition of the material being deposited from metallic to fully stoichiometric compound film. This shows in turn that careful consideration should be given to obtain a deposition condition to allow operation at, or very near, the transition point. Practically, it is difficult to operate near the transition point due to many variables, since a very small change in the flow rate leads to a large change in the partial pressure of the reactive gas which consequently affects the deposition rate. In order to avoid target or evaporant source poisoning, various techniques have been employed such as maintaining a gradient in composition of the reactive gas by injecting argon near the target and the reactive gas near the substrate or by arranging additional gettering surfaces in the chamber to maintain a gradient in composition of the reactive gas (Schiller *et al.*, 1979; Maniv *et al.*,

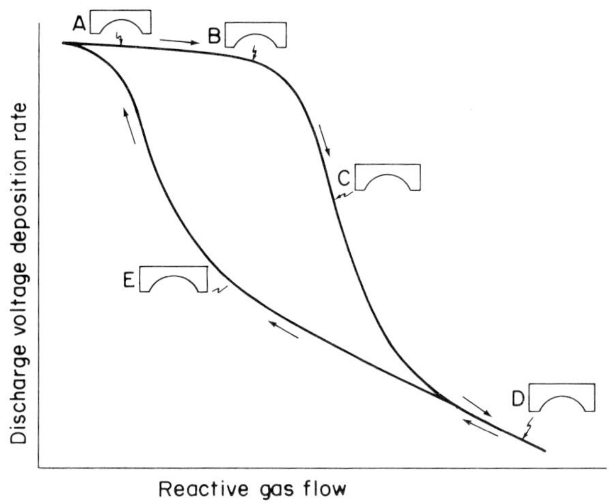

Figure 65. Typical dependence of deposition rate and discharge voltage (at given power and pressure on reactive gas flow during reactive sputtering of metals (from Bunshah and Deshpandey, 1986). (*Reproduced by permission of the American Institute of Physics*)

1983; Schorer and Wirz, 1984). Other techniques employed involved the use of a positively biased anode near the substrate to intensify the plasma adjacent to the substrate and increased the reactivity at the substrate, or the injection of pulsed reactive gas in a controlled manner during the deposition of metals (Sproul, 1984; Ahmed, 1986b).

In general, reactive ion plating has the advantages of depositing thick coatings of various compound phases at low substrate temperatures so that compound phases which usually form at temperatures above 1000°C and cannot normally be used on tool steel without causing softening, can be produced at temperatures as low as 350°C with this technique.

5.2 EXAMPLES OF REACTIVE ION PLATED COATINGS

5.2(a) Nitride Coatings

5.2(a.1) Titanium Nitride

Perhaps the most successful application of reactive ion plating is in the production of titanium nitride coatings because of the ability of the process to deposit such coatings below the tempering temperature of typical hot working and metal cutting tool steels (Matthews, 1985). Titanium nitride coating (TiN) has a faced-centred cubic crystal structure with a high melting point of about 2900°C. The high hardness of titanium nitride, which is around 4000 kgf mm^{-2}, its corrosion resistance and relative inertness as well as its yellow-gold colour makes it attractive in the tool (Buhl *et al.*, 1981), electronic (Wittmer *et al.*, 1981) and decorative industries (Zega *et al.*, 1977). In order to produce titanium nitride coatings for different applications, variable reactive ion plating processes have been utilized including triode assisted discharge (Matthews and Teer, 1979), arc evaporation (Hatto and Teer, 1986), hollow cathode discharge (Sato *et al.*, 1978) and sputter ion plating (Coad and Restall, 1982). However, the observed physical properties of coatings prepared by the above methods can vary over orders of magnitude depending on both the growth technique and the particular deposition parameters employed (Sundgren *et al.*, 1983; Fleischer *et al.*, 1979; Cheavallier *et al.*, 1981). Looking at the above techniques one can notice a common factor shared by all of them and that is their capabilities of increasing the ionization efficiency markedly during the deposition of the metal. Ionization efficiency levels as high as 100% have been reported for evaporant species in the case of arc evaporation (Dorodnov, 1981a,b). As discussed in the previous section, the increase in ionization efficiency can greatly influence crystallographic structure, coating structure, hardness and composition of titanium nitride coatings. Therefore, due to its importance in tribological applications this subject has received great attention in recent years. For example, it has been reported by Matthews and Sundquist (1983) that the increase in ionization efficiency changed the stoichiometry of titanium nitride coating from (TiN) to (Ti_2N) which has a higher hardness than (TiN). However, the ionization efficiency is not the only parameter to control in order to achieve a stoichiometric TiN coating since other factors such as nitrogen partial pressure, deposition rate, substrate temperature and substrate bias have to be optimized as well. In most TiN ion plating processes, mixtures of argon and nitrogen gases are used at certain ratios during the evaporation. The argon gas is usually added to dilute the reactive gas for achieving better control over the stoichiometry of the deposited nitride coating. In general, ratios of nitrogen to argon gas used can vary from 0.01:1 to 6:1 depending on other process parameters (Delcea and Dunare, 1983; Valli, 1984; Komiya *et al.*, 1979; Navinsek and Zabkar, 1986). Although operating at high total pressure has certain advantages to the coatings such as better thickness and discharge uniformities, working at lower total pressure creates the possibility of improving film properties because of the enhancement of the bombarding ion energy through reduced charge exchange collisions and/or by reduced neutral (energetic and thermal) bombardment (Suzuki and Howson, 1983). For this reason r.f. bias ion plating with an electron beam evaporation source is satisfactory in the production of high quality optical titanium nitride coatings (Suzuki and Howson, 1983). The other complex factor which has to be controlled is the ratio of the evaporation rate of titanium to the partial pressure of nitrogen. It has been reported by Jacobson *et al.* (1979) and Matthews and Sundquist (1983) that the decrease in evaporation rate changes the deposition from Ti to Ti_2N to TiN including a two-phase mixture. Figure 66 shows the range of compositions produced across the range of titanium evaporation rates and nitrogen inlet flow rate as carried out in a thermionically assisted triode

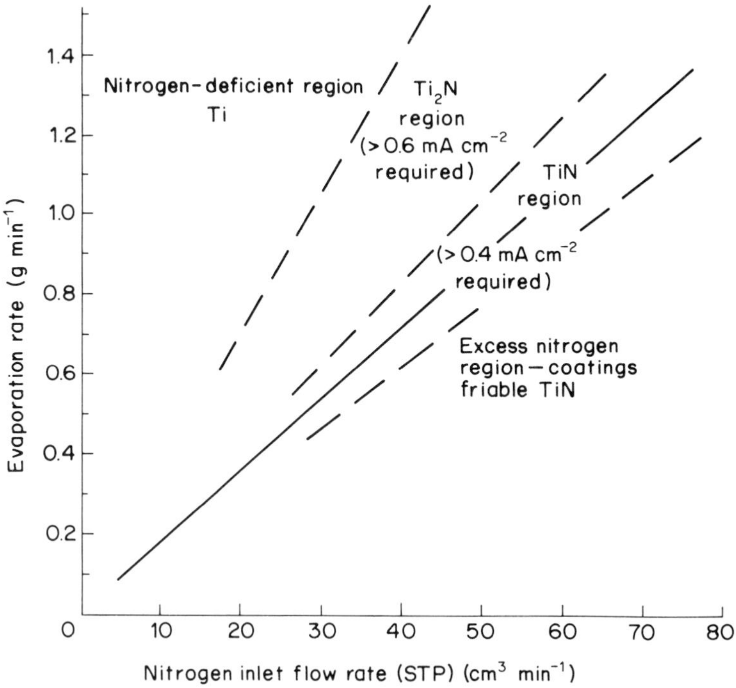

Figure 66. TiN compositions produced across the range of titanium evaporation rates and nitrogen inlet flow rates (from Matthews and Teer, 1980). (*Reproduced by permission of Elsevier Sequoia SA*)

reactive ion plating (Matthews, 1980). This figure shows the interaction between the discharge current density, the nitrogen flow rate and the deposition rate of titanium metal. The change in the nitrogen partial pressure can also influence the colour of the coating. It has been reported by many workers that as the nitrogen level is increased during the coating process, the colour of the titanium nitride coating changes from a light metallic gold at low levels to a yellow gold and then to a brownish gold at high nitrogen levels (Munz *et al.*, 1983; Sproul, 1985). Valli (1984) reported that in r.f. bias ion plating process, if the ratio of nitrogen to argon gas is around 5 then a titanium nitride coating with a gold colour can be obtained at fixed evaporation rate and power density. In many applications, however, a colour variation across the substrate can be observed which might be due to the variation in thickness or power density on the sample or due to the non-uniformity in gas flow or substrate temperature. In fact, substrate temperature can have a great influence on the structure and the adhesion of the coating as well. The main reasons why the coatings have to be deposited on a high temperature substrate in certain applications are to give high reactivity and high surface mobility as thermal energy to the depositing particles (Suzuki and Howson, 1983; Wickersham and Pool, 1985). However, in reactive ion plating, the increase in ionization efficiency can give high reactivity at room temperature because of ionic bombardment, thus eliminating the need to heat the substrate (Matthews, 1985).

The ionic bombardment should be uniform over the entire surface of the coated component to ensure a uniform power input. Therefore, small components will be more prone to overheating because of the high discharge power input over a small area (Matthews, 1985). Overheating to above 400–500°C will affect the adhesion of the titanium nitride coating on the substrate. Helmerrson *et al.* (1986) have reported the formation of a brittle carbide phase at the interfaces where the temperature was raised above 500°C, which reduced the affective adhesion.

5.2(a.2) Silicon Nitride

In recent years there has been considerable interest in the deposition of silicon nitride for its electrical and optical properties. Chemical vapour deposition processes have been widely used to produce such coatings. The method, however, has a disadvantage in regard to its high substrate temperature which is not desired for device fabrication. To meet the requirement of depositing at low substrate temperature, different processes have been utilized including the ion plating technique. A diode type ion plating system has been used to evaporate silicon in a nitrogen or ammonia gas discharge. This system has the disadvantage of low reactivity due to the low ionization efficiency and the high operating pressure (Chin and Elsver, 1974). Films deposited in this method are usually non-stoichiometric Si_3N_4 with high metallic contents. On the other hand, by using an r.f. bias ion plating system a transparent, stoichiometric film of Si_3N_4 have been successfully deposited on substrates at room temperature (Suzuki and Howson, 1983). In this method, silicon 'metal' was evaporated from a water-cooled copper crucible using an electron beam gun in a nitrogen discharge at pressures less than 0.1 Pa. In general, the formation of Si_3N_4 requires a considerable activation of the reactants at low discharge pressure. Therefore, additional plasma activation, by means such as thermionic supported glow discharge or r.f. discharge, enables one to lower the deposition temperature and increase the ionization efficiency (see Section 4.1). In addition to the pressure effect, it has been also observed that the best quality silicon nitride films require a good balance between the evaporation rate and r.f. power (Suzuki and Howson, 1983). In order to separately control the different variable parameters such as deposition rate, nitrogen flux rate and pressure, Weissmantel (1976) used a dual ion beam system to deposit stoichiometric Si_3N_4 by sputtering a Si target using an inert gas ion beam, while the condensing layer was simultaneously bombarded with nitrogen ions supplied from a second ion beam source. Adherent and smooth silicon nitride films were obtained in this process. When ammonia gas is used instead of nitrogen the silicon nitride film will contain trapped hydrogen which can affect some of the electrical properties of the film (Hu *et al.*, 1967).

5.2(b) Oxide Coatings

5.2(b.1) Aluminium Oxide (Al_2O_3)

Aluminium oxide has distinctive electrical, optical and mechanical properties (Gouault *et al.*, 1971). The high dielectric constant of aluminium oxide and its success in providing radiation resistance as well as its ability to provide corrosion protection and wear resistance in gasifier fabrication (Swaroop *et al.*, 1976) have stimulated many programmes to investigate the most efficient method of producing aluminium oxide films. However, the film properties depend upon the preparation techniques and parameters. The ion plating process has been used by different workers for the deposition of stoichiometric aluminium oxide coating. Swaroop *et al.* (1976) deposited alumina in an oxygen-rich discharge and at relatively high evaporation rate. However, they reported a coating hardness of only one-quarter than that of the bulk alumina. The substrate temperature was maintained at 350°C. Bunshah and Schramm (1977) used the activated reactive evaporation process to evaporate aluminium in a partial pressure of oxygen of less than 0.3 Pa. They reported an Al_2O_3 structure at high substrate temperature and evaporation rate. Nakahara and Hebard (1983) described the synthesis of metal oxides by reactive ion beam sputter deposition. They showed that as the partial pressure of oxygen is increased, amorphous and smooth alumina films with high electrical resistance were obtained. The author (Ahmed and Teer, 1982; Ahmed 1986b) used a pulsed oxygen process in a thermionically assisted triode ion plating process to deposit dense, hard and highly adherent aluminium–aluminium oxide coatings. He reported that as the partial oxygen flow rate increases at high substrate bias, a non-crystalline phase of near-stoichiometric Al_2O_3 is obtained (Figures 67 and 68). It is clear from the above that in the ion plating of aluminium oxide coatings a partial pressure of oxygen and argon is used. By keeping the percentage of oxygen down to the minimum required for the complete oxidation of the depositing film, a substantial increase in the deposition rate can be obtained (Deshpandey and Holland, 1982). However, there is a critical partial pressure of oxygen below which oxygen-deficient alumina will be formed with metallic rather than ceramic characteristics. It was reported by Cook (1983) that a partial pressure of oxygen of at least 0.1 Pa is required in an r.f. ion plating system to produce an electrically insulating alumina film. In such a system, however, the oxygen partial pressure and the electron beam current are inextricably linked and careful control of both the electron beam power and gas flow is required to maintain stable evaporation. Furthermore, Cook (1983) reported that it was beneficial to warm up the electron beam source by slowly increasing the current and simultaneously increasing the oxygen partial pressure in the discharge. If the evaporation is established before the oxygen is added then film adhesion is impaired.

5.2(b.2) Titanium Dioxide (TiO_2)

Titanium dioxide has been used mainly for its optical properties of high refractive index and transparency in the visible and IR regions of the light spectrum. In

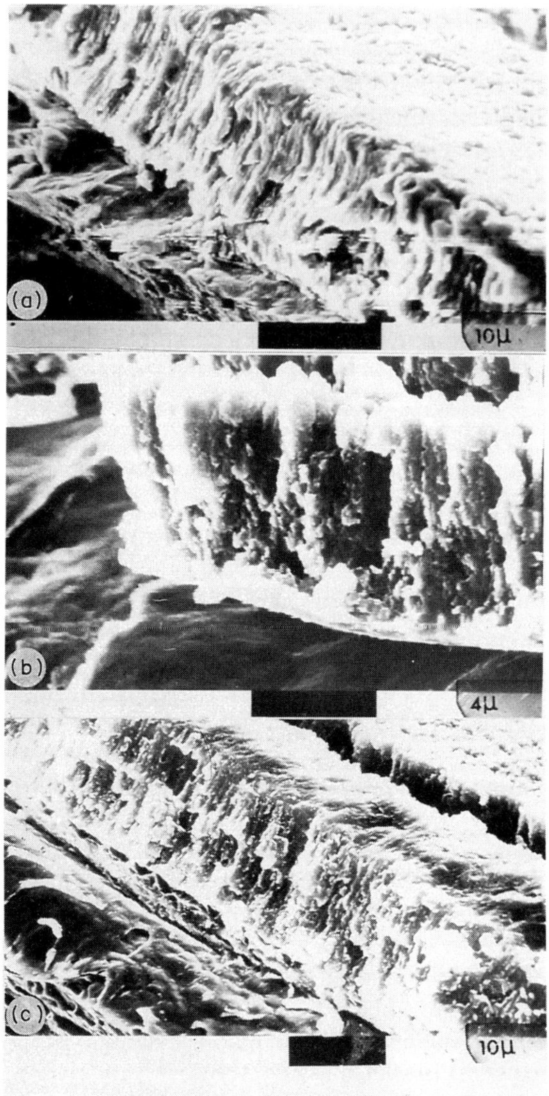

Figure 67. Scanning electron micrographs of fractured sections of aluminium coatings prepared at 0.1 Pa argon pressure, 2 kV bias, oxygen pulse rate of 3 s duration at 3 s intervals and at oxygen flow rates of: (a) 1.5, (b) 4.5, and (c) 7.5 Torr 1 s^{-1} respectively

general, evaporation and sputtering techniques were used to prepare such coatings at partial pressure of oxygen and an elevated substrate temperature (Pulker, 1976; Pawlewiez and Loegreid, 1978; Pawlewiez *et al.* 1982; Fan *et al.*, 1974; Schiller *et al.*, 1980). Sources used included TiO_2, TiO and Ti for

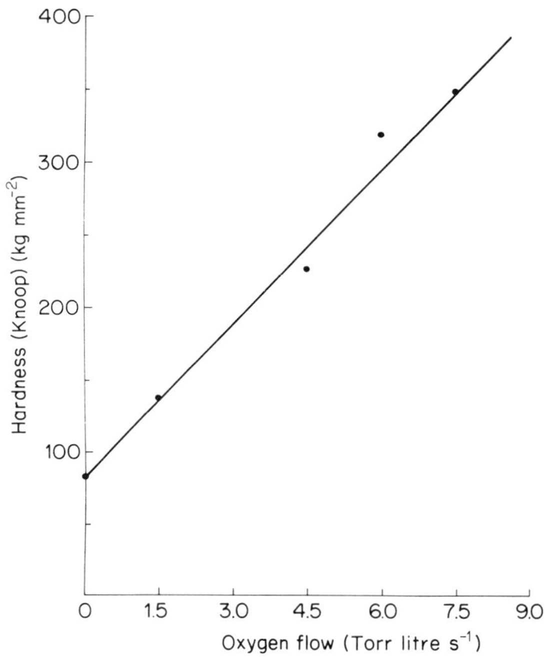

Figure 68. Dependence of hardness on oxygen flow rates for aluminium coatings prepared at 0.1 Pa argon pressure, 2 kV bias and at oxygen pulse rate of 3 s duration at 3 s intervals

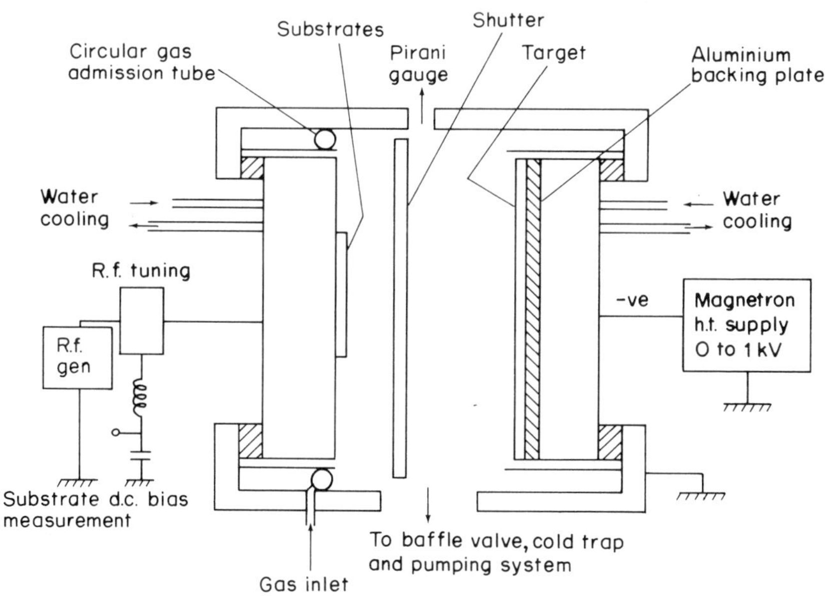

Figure 69. D.C. planar magnetron sputtering with r.f. ion plating (from Howson *et al.*, 1984). (*Reprinted with permission from Vacuum vol. 34, ©1984 Pergamon Journals Ltd*)

evaporation (Pulker, 1976), Ti for d.c. and r.f. sputtering (Schiller *et al.*, 1980; Fan *et al.* 1974) and TiO_2 for r.f. sputtering (Schiller *et al.*, 1980). The disadvantage of these processes, however, was the difficulty of depositing TiO_2 films on plastic substrates at room temperature. In recent years, an r.f. bias ion plating system using electron beam evaporation source has been utilized to deposit high refractive index TiO_2 films onto insulated substrates at room temperatures (Howson *et al.*, 1984) (Figure 69). Titanium metal was evaporated in an atmosphere of argon and oxygen. In order to obtain films equivalent to those produced at higher substrate temperatures, the evaporation rate of titanium has to be balanced carefully with the partial pressure of oxygen and the r.f. power applied on the substrate.

Chapter 6
The Characteristics of Ion Plating Coatings

6.1 ADHESION

The durability of coatings is usually of prime importance in many applications and it is largely dependant on the stress and therefore on the adhesion between the coating and the substrate. This is particularly important if the coating or substrate is to be exposed to harsh environment such as corrosion, high temperatures, high forces and humidity (Green and Zilko, 1978). In such cases the requirement for good adhesion means that the interfacial region between the coating and the substrate should not fall under different test conditions (Mattox, 1982b). In general, good adhesion is usually promoted by the strong atom–atom bonding within the interfacial region, low local stress levels, the absence of easy deformation or fracture modes and the absence of long term degradation modes (Mattox, 1978; 1982b). These characteristics depend largely on the nature of the interfacial region, which in turn depends on the interactions between the depositing material and the interface. In ion plating, excellent adhesion is usually obtained even when the coating and substrate do not alloy and when diffusion is difficult (Mattox, 1973; 1980; Spalvins, 1985). Comparison between the adhesion of films produced by ion plating and other coating techniques have always indicated the superiority of the ion plated films in this respect (Spalvins *et al.*, 1966; Mattox, 1973; Teer, 1977). Figure 70 shows a comparison between the adhesion of vacuum deposited silver film and ion plated silver film on mild steel tensile specimens which have been pulled to failure (Teer, 1983). The coating produced by vacuum evaporation has become detached completely whereas the ion plated coating was still adherent up to the point of fracture. Spalvins *et al.* (1966) reported an excellent adhesion of ion plated coating even in the case of incompatible materials. He reported a graded interface of 6 μm in thickness for gold on nickel and related that to the continuous solid solubility between the two metals (Figure 71). Meanwhile, for materials which have very limited solubility such as gold on iron, he reported an interface of approximately 1.5 μm in thickness (Figure 72). On the other hand, gold vacuum deposited on nickel and steel exhibited an abrupt interface as shown in Figure 73. Teer (1983) has demonstrated that when a wide variety of ion plated coating/substrate combinations were subjected to bend, tensile, shear and scratch test, thermal shock and ultrasonic vibration, no failure of the adhesion was detected. Similarly, Mattox (1965) has listed 19 different metal/metal combinations for which he reported excellent adhesion, and metal-on-semiconductor pairs where a wire soldered to the metal and subsequently pulled caused failure in the solder and not at the interface. Ion plated coatings produced

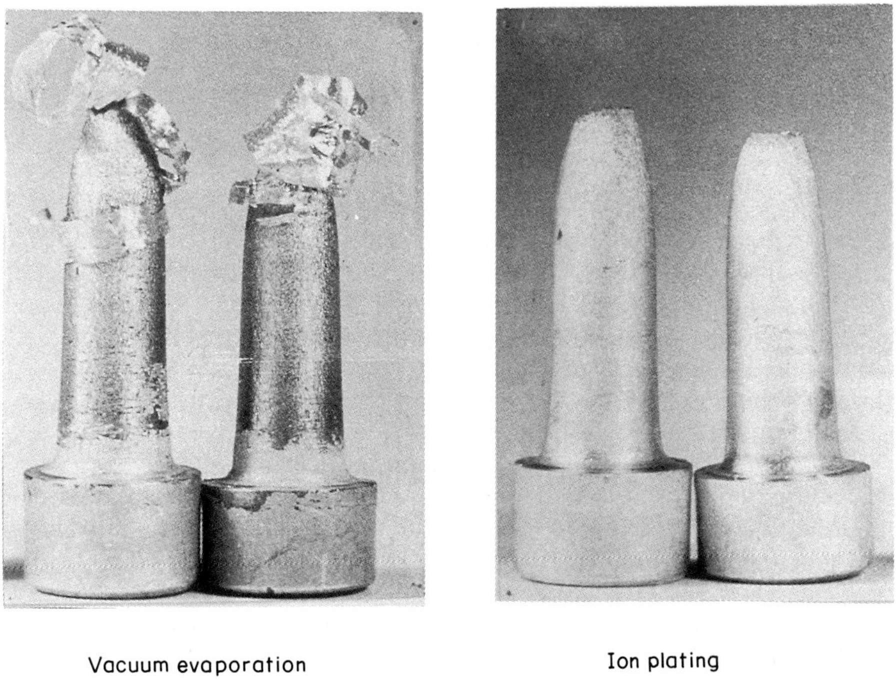

Figure 70. Silver on steel tensile specimens, ion plated on the right and vacuum evaporated on the left (from Teer, 1983). (*Reproduced by permission of EEC, Brussels-Luxembourg*)

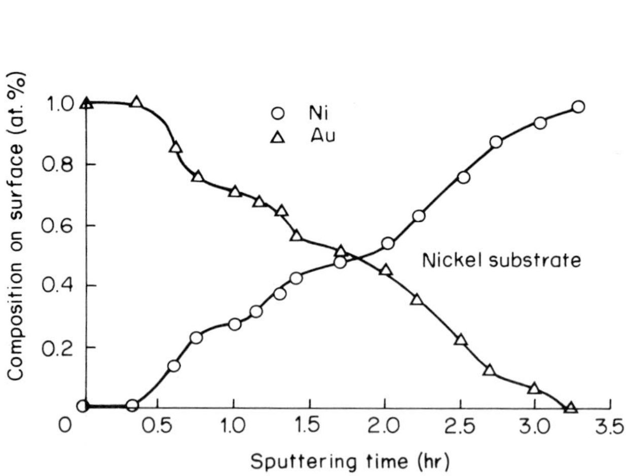

Figure 71. Depth profile of 0.6 μm thick gold ion plated on nickel (T Spalvins, Private Communication, 1985)

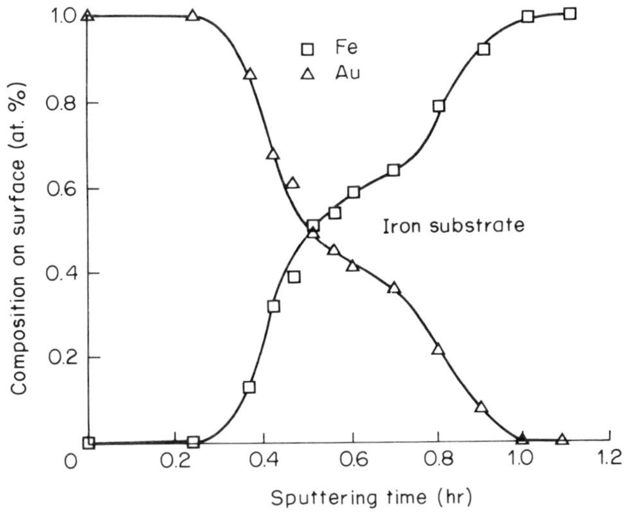

Figure 72. Depth profile of 0.6 μm thick gold ion plated on iron (T Spalvins, Private Communication, 1985)

by the author on different substrates have never failed even when the samples were pulled to fracture (Ahmed and Teer, 1981; 1984a,b).

It is generally believed that the formation of a graded interface is mainly responsible for the excellent adhesion of ion plated coatings on various substrates. The mechanism of the formation of such an interface has been discussed in detail in Section (2.8a). However, while it is possible to understand the good adhesion at the interface between mutually compatible materials, it is rather difficult to understand such a mechanism in the case of incompatible pairs where the conventional diffusion is limited, and the interface is shallower (Teer, 1983). As it has been pointed out in Section (2.8a), the low energies of incident particles on the substrate will lead to penetration depths of few tens of Ångströms which are sufficient to account for the injection of coating particles into the substrate and into the growing coating but not transferring substrate atoms into the coating to provide broad graded interface. However, one such possible mechanism is radiation enhanced diffusion, where diffusion is enhanced by the creation of lattice defects via ion bombardment thus promoting deep penetration of the coating into the substrate (See Section 2.8a).

It has also been reported that the adhesion of incompatible materials can be improved substantially by low and high energy ion bombardment (Ahmed, 1985a; Jacobsson *et al.*, 1983). For example, Franks *et al.*, (1979) modified the adhesion of gold to various substrates with 5 keV argon ion bombardment. Salem and Sequeda (1981) observed an increased adhesion of gold to silicon and glass when the substrates were pre-irradiated with oxygen ions. Many other workers have also reported remarkable improvements in gold adhesion for oxygen-ion assisted

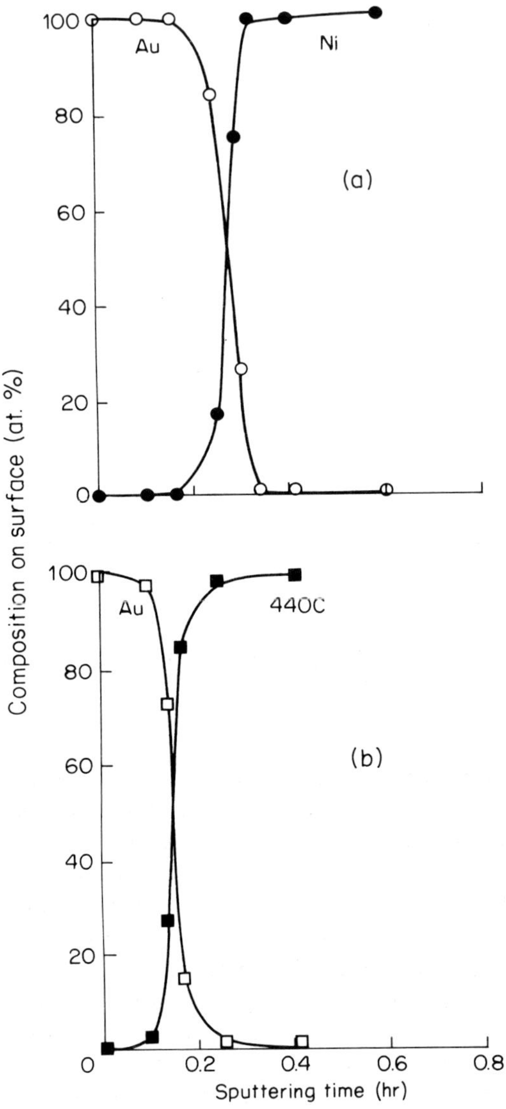

Figure 73. Depth profile of 0.3 μm thick gold vacuum deposited on copper, (a) nickel and (b) 440 C stainless steel (T Spalvins, Private Communication, 1985)

deposition. The author has found that the adhesion of gold on silica increased substantially when oxygen was used to sputter the gold film on the silica substrate followed by ion bombardment with 10 keV Ar^+ ions (Figure 74) (Ahmed, 1986a,c). Furthermore, Martin *et al.* (1984) have found that the scratch test failure

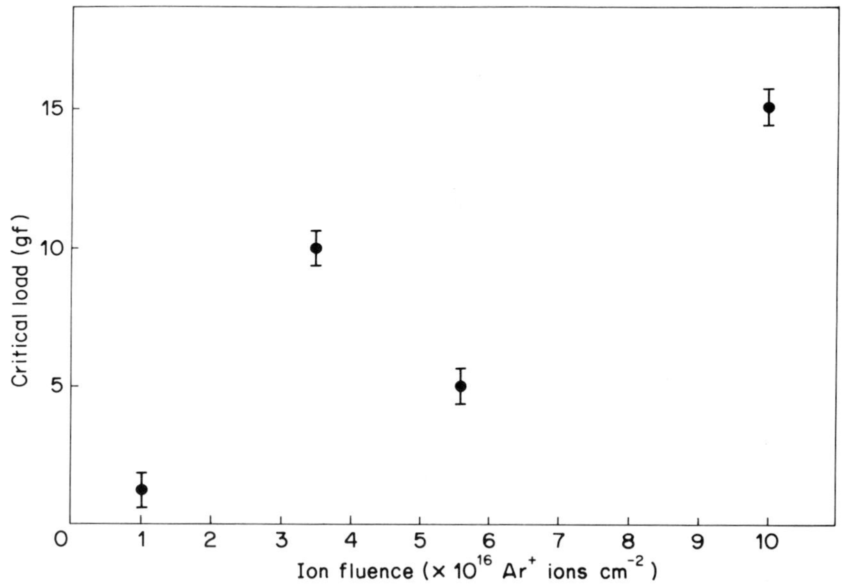

Figure 74. Scratch test results of dynamic recoil mixed gold on silica using 10 keV ar^+ ion beams at different Ar^+ ion fluences (N. A. G. Ahmed, unpublished results)

loadings for gold films were raised by a factor of 1000 before film removal from glass and silica substrates after ion beam treatment with oxygen. On the other hand, a substantial increase in adhesion has also been observed followed the irradiation of the coating with electrons of energies from 5 to 30 keV (Mitchell *et al.*, 1984). Although no theory on the origin of such a bonding process has yet been established, it is generally beleived that the enhancement of adhesion following low energy ion bombardment is mainly due to nuclear interactions produced by the passage of ions through the interface. When oxygen is used, the improvement in film adhesion can be attributed to the reduction of the porous interfacial layer between the coalescent film and the substrate during the initial growth of the film and possible formation of gold oxide in the interface (Martin, 1986). Therefore, in ion plating some of these processes may also contribute to the increase in adhesion.

6.1(a) Adhesion Testing of Thin Hard Coatings

Hard coatings such as TiC, TiN, HfN and χ–Al_2O_3 are widely used now in the tool industry to increase the wear resistance and the speed of cutting tools (Hatschek, 1983). In these applications, however, the adhesion of the hard coating to the substrate is of vital importance for the performance of the coated tools since high forces are involved in the cutting process (Hatschek, 1983). Therefore, it is important to test the adhesion during the quality control stage in the production of large numbers of parts to ensure consistent results in practical applications. The adhesive strengths of the hard coating to the substrate may vary from 100 to 3000 N mm^{-2}. However, for adhesive strengths of less than 100 N mm^{-2} only a few methods have been reported for the adhesion testing of hard coatings on various substrates (Perry, 1983). Among the various techniques proposed by different workers for testing the adhesion of thin hard coatings, the scratch test has been shown to be the only test that produces consistent results (Helmersson *et al.*, 1985).

The scratch test was originally proposed by Heavens (1950), who used a smoothly rounded chrome-steel point which was drawn across the film surface. The method was put on a quantitative basis by Benjamin and Weaver (1960) for testing the adhesion of soft coatings. However, for testing the adhesion of hard coatings, Perry and Archer (1980) suggested a slightly different approach by using the scratch test to introduce stresses at the film–substrate interface.

The stresses are induced by deforming the coating surface by means of the indentation of a moving diamond tip 0.1–0.2 mm in radius and 120° wedge (Figure 75). The applied load is increased until the deformation causes stresses which result in flaking (adhesive failure) or chipping (cohesive failure) of the coating. The load can be increased either stepwise or continuously. The smallest load at which the coating is stripped from the substrate either by adhesive or cohesive failure is termed the critical load and is determined by optical or electron microscopy and/or by acoustic emission (AE). In the latter technique, an acoustic signal detector in the form of an accelerometer is mounted just above the diamond stylus to register the scratch signal emitted in the range 0–50 kHz or 0–200 kHz (Perry, 1983). The acoustic signal is then plotted on a chart recorder as a function of the stylus load. When adhesion failure occurs, the signal can increase either suddenly or steadily as a function of load (Figure 76). Perry (1983) has used this method to evaluate the adhesion of TiC and TiN films deposited on high speed steels. He concluded that in spite of the difficulties in calculating adhesion from the scratch test results, due to the sensitivity of the method for substrate and film–bulk properties, the test gives a value of the toughness of the film–substrate composite where the adhesion is only one of the important parameters. In a further study, Hintermann (1984) found a good correlation between the onset of the AE signal and the microscopical observation of the first damage occurring. He also reported that the scatter of repeated determinations of the critical load on the same specimen is generally better than 10%.

It can be concluded, therefore, that adhesion measurements carried out by the scratch test give qualitative information which is good enough for comparison between different samples if used with caution. The results, however, depend on many factors such as the coefficient of friction between the scratching diamond and the film, mechanical properties of the film and the substrate, and the thickness of the film (Helmersson *et al.*, 1985).

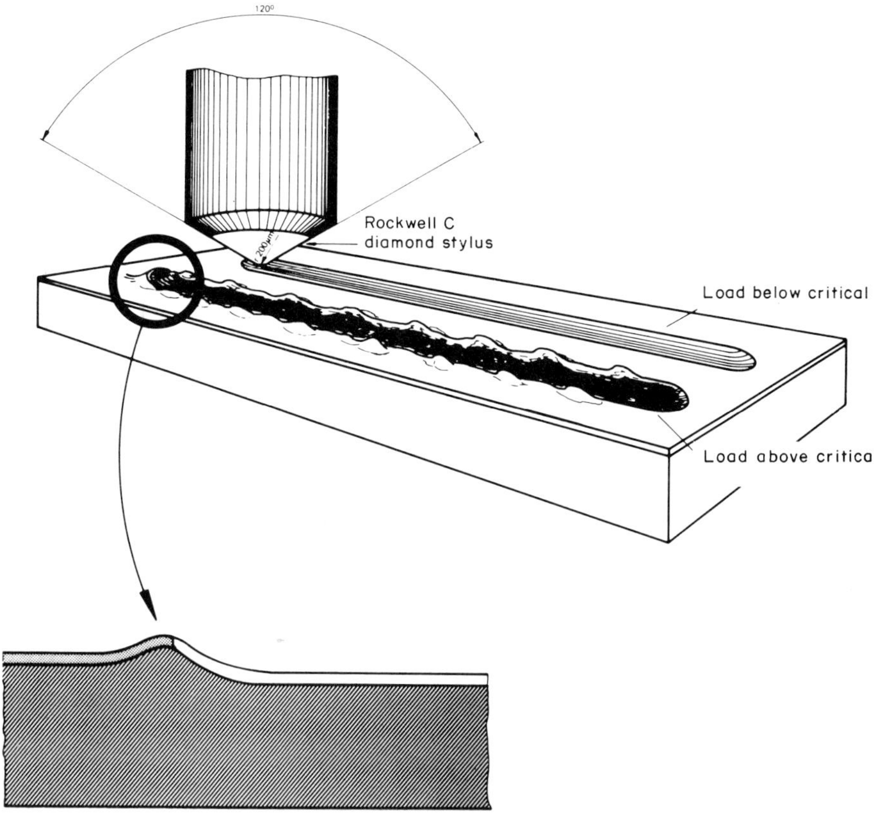

Figure 75. Schematic representation of the diamond stylus and the form of channels below and above the critical load in the scratch testing of hard coatings (from Perry, 1983). (*Reproduced by permission of Elsevier Sequoia SA*)

6.2 THICKNESS UNIFORMITY

Another advantage of the ion plating process is its ability to produce a uniform thickness on all sides of a stationary substrate. Unlike vacuum evaporation, ion plating has shown a fairly uniform coating thickness on surfaces of complex shapes without rotation of the substrate (Figure 77). This property of ion plating is often called 'throwing power'. Scattering of the metal vapour through numerous collisions in the path from the vapour source to the substrate is believed to be the main reason for this thickness uniformity. This effect has also been seen at high gas pressures without the discharge present, but ion bombardment is important in obtaining adherent coatings since evaporation into a high gas pressure environment will lead to vapour phase nucleation and the deposition of a fine powder as employed in 'gas evaporation'. Chambers and Carmichael (1971) and Wan *et al.* (1971, 1973) have studied the thickness

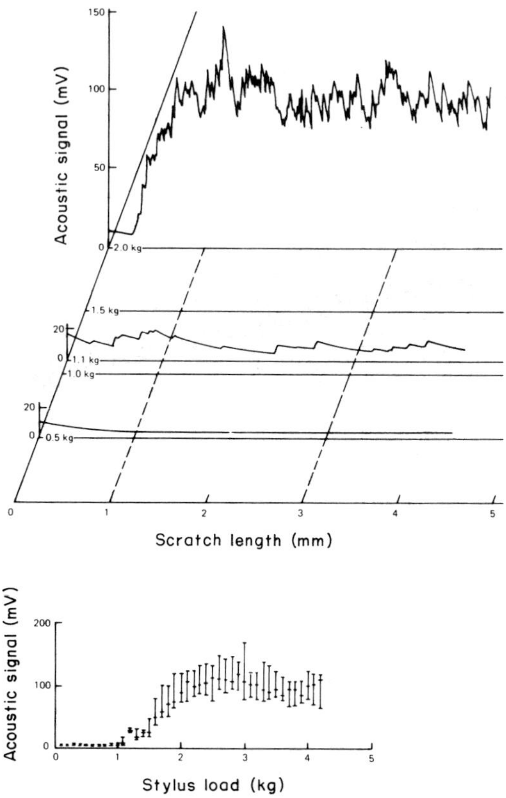

Figure 76. Acoustic signals recorded in the 0–50 kHz range when stainless steel coated with TiN by ion plating was tested (from Perry, 1983). (*Reproduced by permission of Elsevier Sequoia SA*)

variations of ion plated gold on various stationary shapes, i.e. flat surfaces as well as solid and hollow cylinders. They showed that the coating deposition rate on the front and back surfaces of a flat plate, solid cylinder and internal surfaces of a hollow cylinder depends on the pressure of argon gas in the vacuum chamber at fixed source-to-substrate distance. Furthermore, they also showed that the ratio of the front to back surface thickness (R) decreases to unity as the argon sputtering pressure increases. On the other hand, Kennedy *et al.* (1971) found that uniform coating thicknesses can be achieved at high condensation rates on substrates having no potential or a negative potential relative to the evaporation source and that the thickness ratio of the front to the back (R) is strongly dependent on the pressure in the plating chamber.

Teer (1975), Sherbiney and Aboukhashaba (1975) and the author have studied the throwing power of ion plating on flat specimens at different gas pressures

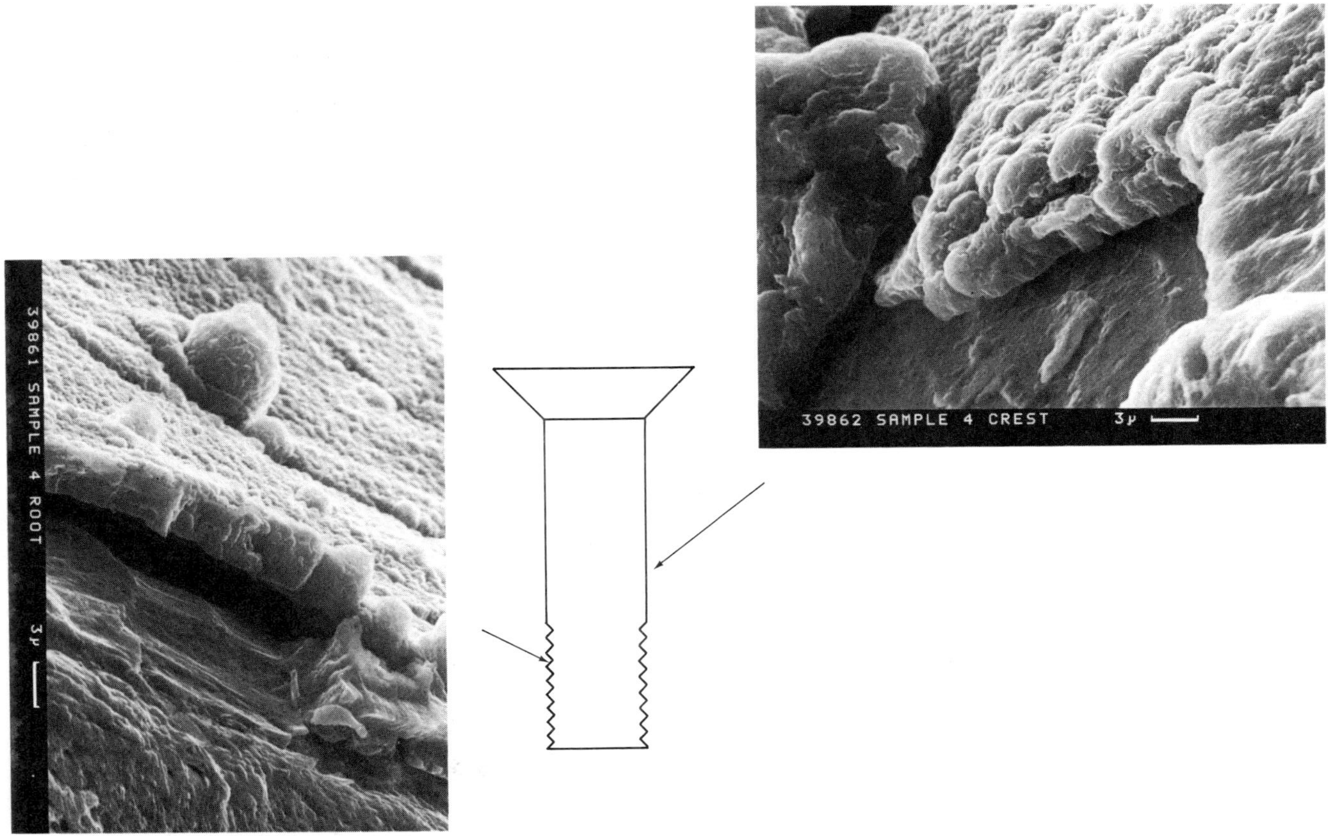

Figure 77. Thickness uniformity of ion plated steel fasteners with aluminium without rotation (N. A. G. Ahmed, unpublished work)

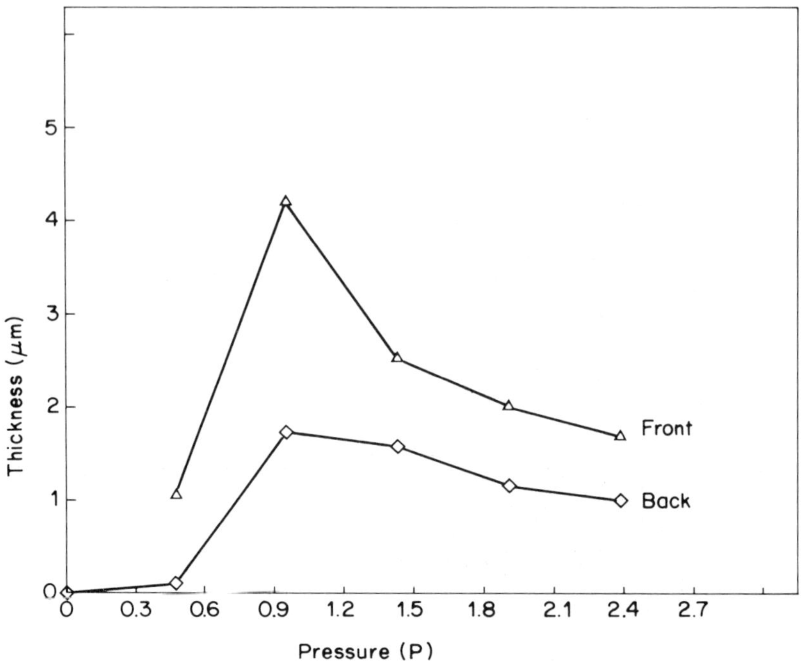

Figure 78. Effect of argon discharge pressure on coating thickness on back and front surfaces of a flat plate

and concluded that the throwing power depends mainly on gas scattering rather than bias voltage. Figure 78 shows the results obtained by the author on the effect of argon gas pressure on the coating thickness of the front and back surfaces of a flat sample. Similar results have also been obtained by other workers (Spalvins, 1980, 1981; Fancey and Begnon, 1984) and indicate in general that over the argon pressure range (0.3–0.9 Pa) the thickness on the front and back surfaces increases rapidly, whilst at higher pressures there is a slow decrease in the thickness.

However, while the decrease at high argon pressures can be explained by sputtering and gas scattering effects, no explanation has yet been given for the initial increase in thickness at low pressures. It is possible that such an initial increase is associated with the formation at the lower end of the pressure range (0.3 Pa) of a significant number of negatively charged agglomerates (Aisenberg, 1972; Aisenberg and Chabot, 1973) of film material in the plasma which cannot reach the cathode. The reason for the negative charging of the agglomerates is that in the plasma, the numbers of electrons and ions are very nearly the same and the electron mass is much smaller than the ion mass. Also, the mobility will be different so the random electron particle current density is greater than that for the ion. If an isolated surface is introduced in the plasma, it will charge up negatively to a potential where enough of the incident electrons are repelled so that the net ion and electron current is zero (Ahmed, 1974; Glang *et al.*, 1970). As the pressure increases the agglomerates will be ionized by electron–atom collision and the positive evaporant ions will be accelerated towards the substrate, to add to the neutral bombardment.

On a microscopic scale, the enhanced surface coverage is probably due to sputtering of the deposited atoms into regions out of line-of-sight of the source. Back-sputtering has been used to preferentially deposit materials on vertical side walls even to the extent that all deposited material on the surfaces normal to the impinging ion beam is completely removed (Mattox, 1979).

In general, the 'good throwing power' obtained in ion plating which indicate that the ratio of film thickness on the front surface to that of the back surface (R) of a sample approaches unity at the higher argon pressures, has been utilized in practice to coat three-dimensional surfaces, internal surfaces of tubes and screw heads without specimen rotation (Spalvins, 1981).

Due to the importance of R in the ion plating process control an attempt has been made by Fancey and Beynon (1984) to derive an expression of R in terms of other variables such as gas pressure, substrate area and source-to-substrate distance. By assuming that R is independent of substrate bias voltage, substrate geometry, deposition rate and relative dimensions of the substrate and deposition chamber, they achieved the following expression:

$$R = 150U_R - 0.6 \qquad \text{for } P > 1.0 \text{ Pa} \qquad (6.2)$$

where $U_R = (D\ A^{-1/2} + PS^3\ A^{-1})$, D is a constant of proportionality, A is substrate area, P is the gas pressure and S is the source-to-substrate distance.

However, since high substrate biases may have a great influence on the sputtering of the deposited coating on the front and back surfaces and since a high deposition rate may increase the thickness of the front surface due to the cosine distribution of evaporant material (Holland, 1965), the above expression is of an empirical rather than analytical nature.

When coating complex parts and internal surfaces of tubes, care should be taken not to increase the gas pressure to above 3 Pa in order to increase the throwing power, since a porous structure with a bad adhesion may result. Consequently, the coating will fail rapidly in any practical application such as corrosion protection or wear resistance.

6.3 COATING STRUCTURE

As discussed in Section 2.8(f), thin film condensation process in vacuum evaporation occurs in several different stages. First, very small, discrete, three-dimensional nuclei form at the low energy sites of the substrate surface. The individual nuclei grow, without any appreciable increase in numbers, until neighbouring nuclei begin to link to form a continuous film (Lewis and Campbell, 1967). With further thickness growth, a columnar microstructure can often be observed and thicker film may develop a texture (Figure 79). This

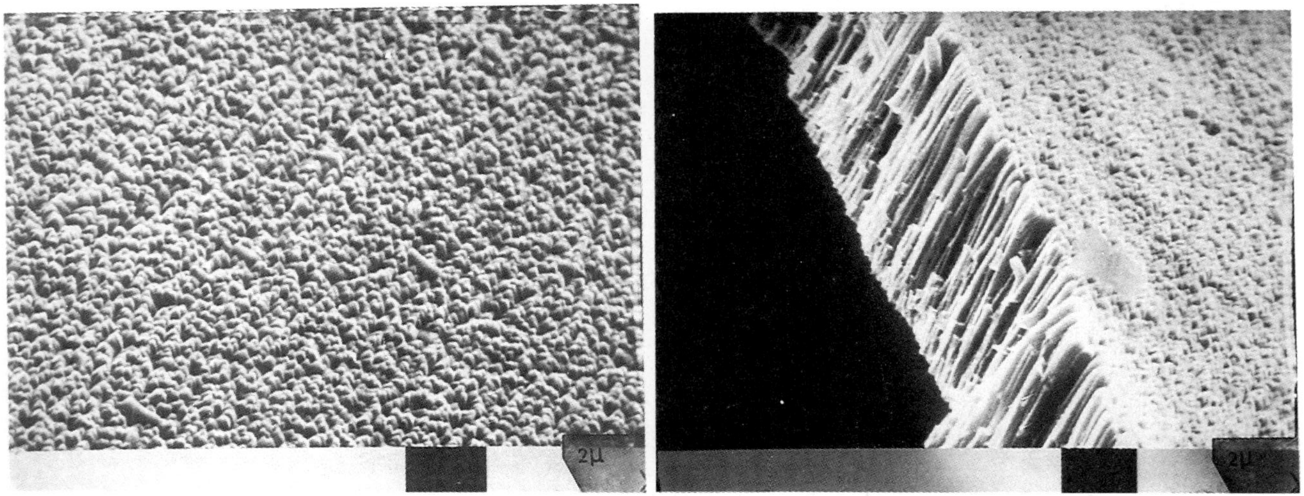

(a)

Figure 79. (a) Top and fractured section of vacuum evaporated aluminium coating

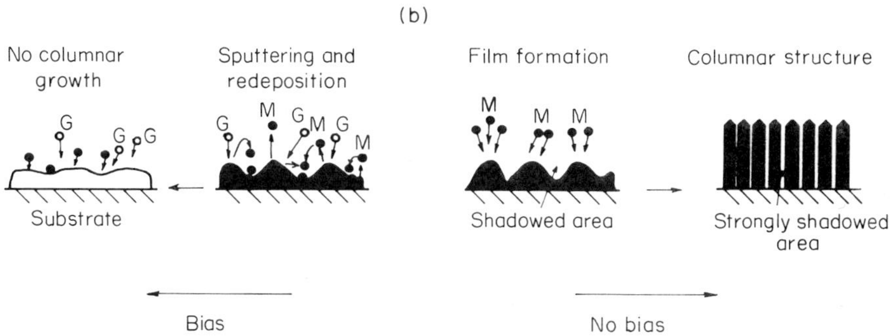

Figure 79. (b) Schematic representation of the microstructural modification due to ion bombardment

process, however, is controlled by many important parameters such as the nature and roughness of the surface, the temperature of the substrate during deposition, the deposition rate, the angle of incidence of the vapour stream and the gas pressure. For example, on a smooth surface, preferential growth of one area over another may result from varying surface mobilities of the depositing atoms as a result of grain orientations. Preferential growth leads to a dominant grain orientation and a surface roughening with thickness. As the surface gets rough, geometrical shadowing will lead to preferential growth of the elevated regions giving a columnar morphology to the deposit (Dirks and Leamy, 1977). An elevated substrate temperature will affect the morphology by increasing surface mobility, enhancing bulk diffusion and allowing recrystallization to occur (Vincett *et al.*, 1977; Wieder, 1970). These effects have led Movchan and Demchishin (1969) to develop a structure zone model of deposited materials at different temperatures. The model, which is shown in Figure 80, involves the formation of three zones, which are determined by the ratio of the substrate surface temperature (T) to the melting point of the deposited material (T_m). Zone (1) occurs at $T/T_m < 0.3$ and results when adatom diffusion is insufficient to overcome the effects of shadowing, giving a columnar structure with low density boundaries between the columns. The individual columns are polycrystalline and are usually highly defected with small grain size. The surface morphology is generally rounded. Zone (2) occurs in the range ($0.3 < T/T_m < 0.5$) where the adatom diffusion dominates and the columnar structure consists of less defected and larger grains with higher density boundaries between the columns. The surface morphology of zone (2) material is generally more regular than that of the zone (1) material. The surface has a smooth, matt appearance. Zone (3) occurs in the range ($0.5 < T/T_m < 1$) and consists of equi-axed grains with a bright surface. Bulk diffusion and recrystallization dominate in this zone. The structure and properties in this zone correspond to a fully annealed metal.

On the other hand, Sanders (1971) suggested a five range of classification based on dominant physical processes. The three lower ranges are similar to

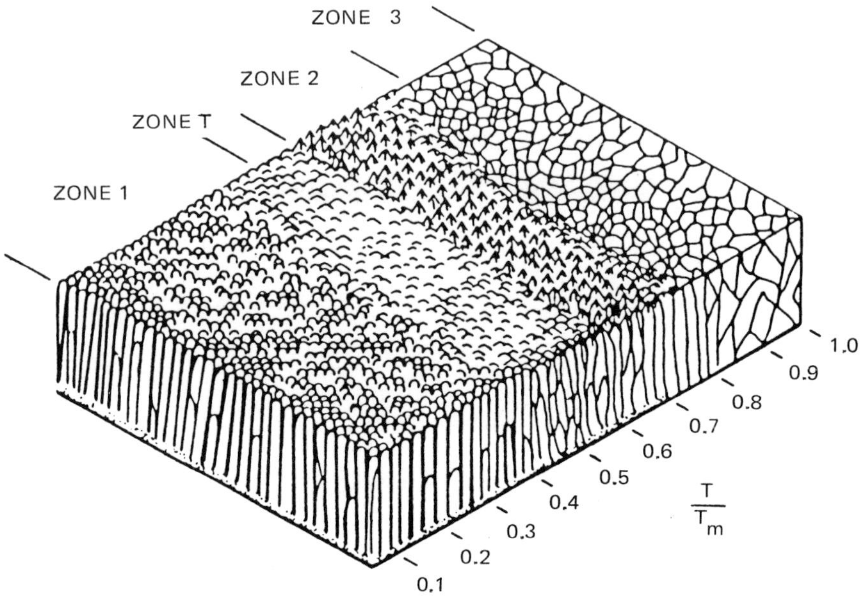

Figure 80. Structural zone model for coating growth as proposed by Movchan and Demchishin (1969)

those of Movchan and Demchishin; in range (1) ($T/T_m < 0.1$), shadowing dominates; in range (2) ($0.1 < T/T_m < 0.3$) surface diffusion dominates; in range (3) ($0.3 < T/T_m < 1$) bulk diffusion is significant. The higher T/T_m ranges refer to liquid condensation.

In a further study, Thornton (1974, 1977) added an additional axis to account for the sputtering argon gas in sputter deposition (Vincett *et al.*, 1977). Thornton showed the effects of both temperature and argon pressure on a three-dimensional diagram (Figure 81) and introduced a new zone (T) between zones (1) and (2). Thus the pronounced zone (1) structure is promoted by substrate roughness, high argon pressure and an oblique component to the deposition flux. The transition zone (T) consists of a dense array of poorly defined fibrous grains without voided boundaries. Zone (2) consists of columnar grains having highly faceted surfaces. At high T/T_m the surfaces are smooth, with grooved grain boundaries. Zone (3) consists of recrystallized grain structure and is characterized by bulk diffusion processes.

Continuous ion bombardment of the growing film with high energy charges or neutral atoms or molecules has been shown to cause changes in the film structure and result in density values near those of the compact material (Takagi, 1982). Studies by Dobrev and Marinov (1973) have shown that ion bombardment of silver, gold, cadmium and cobalt with 1–10 keV argon ions increased the surface mobility of adatoms and accelerated nucleation. Babaev *et al.* (1976) have found similar effects for Ar and Ne ion assisted deposition of Zn on Cu_2O

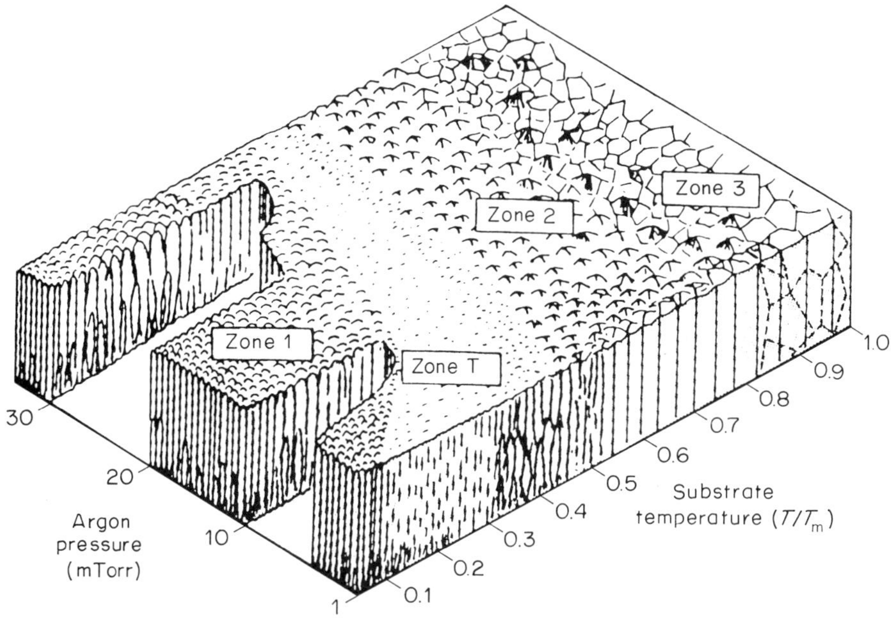

Figure 81. Structural zone model for coating growth as proposed by Thornton for sputtered coatings (from Thornton, 1974). (*Reproduced by permission of the American Institute of Physics*)

and Sb on NaCl. In general, the effect of ion bombardment seems to increase the nucleation density by causing surface damage and to enhance the mobility of adatoms by transferring kinetic energy to them. This effect is important in accelerating the nucleation, growth and coalescence of the nuclei (Gilbert *et al.*, 1978). Furthermore, a significant increase in crystalline perfection of the film can also be achieved if self-ion bombardment (Bland *et al.*, 1974; Bunshah and Juntz, 1972) is present through partial ionization of the evaporant flux and is a promising method for epitaxial film growth (Bland *et al.*, 1974). Ion bombardment during deposition has also been found to suppress the porous boundaries normally found in coatings formed at low substrate temperatures (Mattox and Kominiak, 1972; Vossen and Kern, 1978) (zone 1), and results in a structure similar to zone (T). Such an effect may be caused by the creation of activation sites that stimulate nucleation for the arriving coating atoms, or by eroding surface roughness peaks and redistributing material into valleys. Therefore, instead of the normal prismatic growth, a very densely packed spiky structure of a fine grained crystalline, isotropic films are obtained (White, 1974). Thornton (1977) has conducted a systematic investigation of bias sputtered Cu at low T/T_m and found that the suppression of open boundaries required an ion bombardment flux which is adequate to sputter a significant fraction of the incident coating flux (i.e. 30–60%) and that this fraction increases with the magnitude of the substrate surface irregularities.

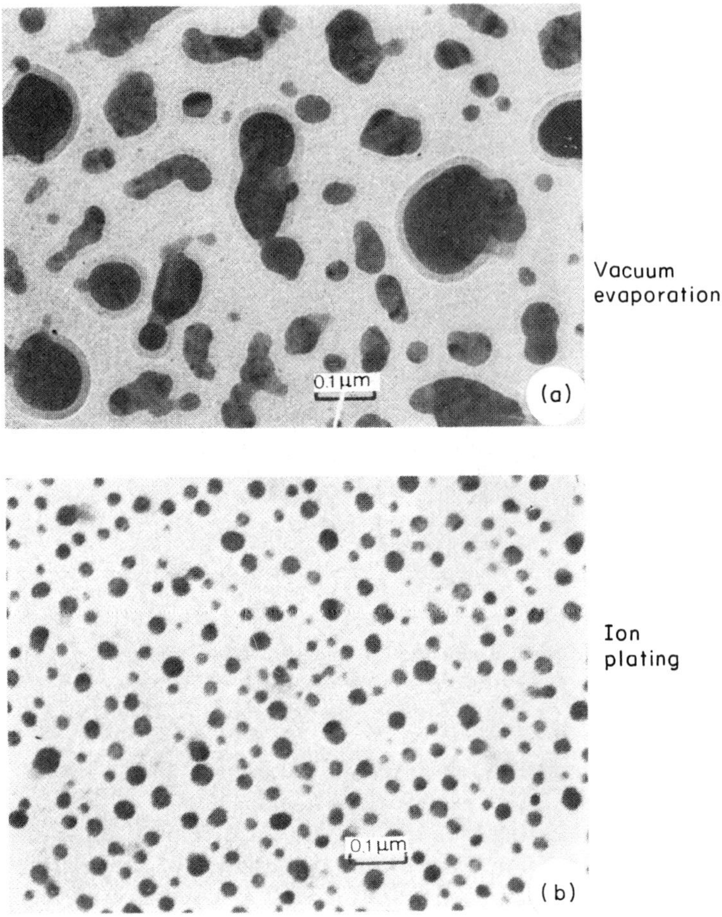

Figure 82. Film nucleation (a) vacuum evaporation, (b) ion plating (from Teer, 1983). (*Reproduced by permission of EEC, Brussels-Luxembourg*)

In diode ion plating it has been observed that the effect of increasing the substrate bias on the coating structure is similar to the effect of increasing the substrate temperature and that the increase in bias voltages produce similar coating structure at lower temperatures (Teer, 1983). Figure 82 shows a comparison between ion plated and vacuum deposited gold film. The ion plated film has a fine, uniform crystalline structure while the vacuum deposited film exhibits a high degree of non-uniformity in the crystallites size and shape (Teer, 1983).

However, although diode ion plating generally gives improved grain structure, such a structure still tends to be columnar as the coating thickness increases (Spalvins, 1981). In order to densify the coating structure at low substrate temperature or at low gas pressures or to deposit high melting point materials, the triode system can be utilized to enhance the discharge (see Section 4.1a). It has been found that with such an enhancement, the coating structure changes from coarse columnar (zone 1) through fine columnar (zone 2) finally to a dense equi-axed structure (zone 3) at very high discharge current density (White, 1974; Salama, 1983). Thus, the formation of coarse columnar structure can be associated with higher glow discharge pressures and low current densities, while the dense equi-axed structure is associated with lower gas pressures and high current densities. Figure 83 shows the structure of ion plated aluminium on steel

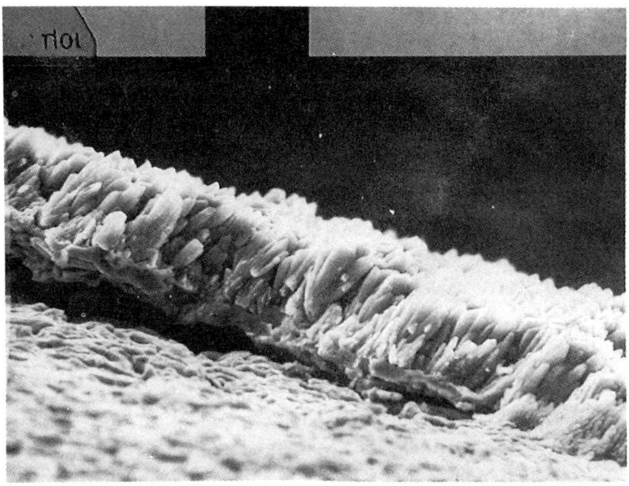

2kV; 1.33 Pa Ar; Diode ion plating; $j = 0.1\ \mu A\ mm^{-2}$

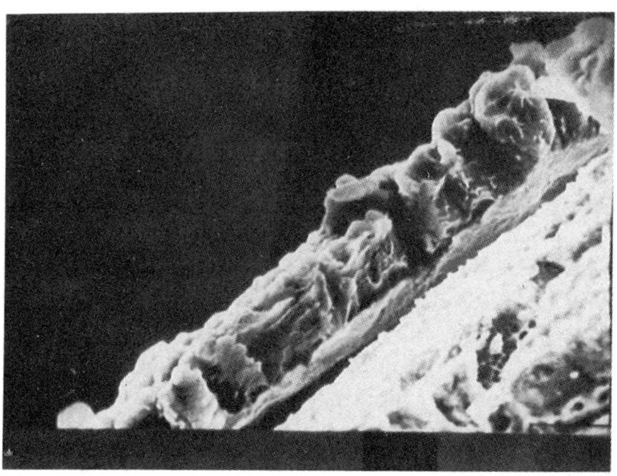

2kV; 1.33 Pa Ar; Triode ion plating; $j = 0.5\ \mu A\ mm^{-2}$

Figure 83. The effect of discharge enhancement on the structure of ion plated aluminium coating

substrate at different current densities. As it has been stated previously, the densification of the structure in the triode system is due to particle bombardment during film growth which increases surface diffusion and changes nucleation and, preferentially, sputters geometric projecting regions, such as tips (Spalvins, 1985).

Therefore, one of the advantages of triode ion plating is that the deposition of high melting point materials such as carbides, nitrides, oxides, etc. can be carried out in the reactive mode on steels and tool steels at low temperatures and without affecting the mechanical properties of such specimens. Larden *et al.* (1978) reported that the temperatures necessary to cause a change from zone (1) to zone (3) for titanium and aluminium coating by ion plating were reduced progressively as the substrate bias was increased, thus increasing the current density and the ion energy. The use of a triode discharge to improve film structures has also been applied by Hecht and Mullaly (1975) in the sputtering of Cu–0.15 Zr and by Matthews and Teer (1979) in the ion plating of TiN. The former workers found that a bias of only -25 V was necessary to produce dense, non-voided structures in a hollow cathode configuration. Such a result is significant because zirconium normally decreases the mobility of the adatoms and the hollow cathode geometry accentuates shadowing effects and also leads to decreased surface mobility, which makes it difficult to obtain dense deposits for this particular alloy (Hecht and Mullaly, 1975). Therefore, by utilizing the enhanced discharge it is possible to achieve dense structure at low working gas pressure.

The improvement of the coating structure following the increase of substrate bias, which affects the energy of the ions and neutrals as well as the current density, led Wan *et al.* (1971) to introduce a relative scale of 'figure of merit of morphology' which is defined as shown in Figure 84. The figure of merit is based on the grain shape, size, surface morphology and density to describe the relationship between the process variables and the microstructure of the ion plating coatings. They found that as the deposition rate is increased, the figure of merit decreases. If the deposition rate is held constant, the figure of merit decreases as the pressure increases as the substrate power density increases.

6.4 WEAR

The wear of components that are subjected to a mechanical action is considered to be one of the main factors responsible for the increase in production costs in industry (Wright, 1981). For this reason considerable efforts are being conducted worldwide to eliminate wear and to increase the performance and lifetime of components. Wear can be generally defined as the removal of materials from one or two surfaces when one surface is moved tangentially with respect to another (Halling, 1979). Therefore, wear can be classified as abrasive, adhesive and corrosive wear, as well as surface fatigue and fretting which occurs as a result of the combination of these classes (Krageliski, 1965). An example of abrasive wear is in metal cutting where debris with high hardness may cause

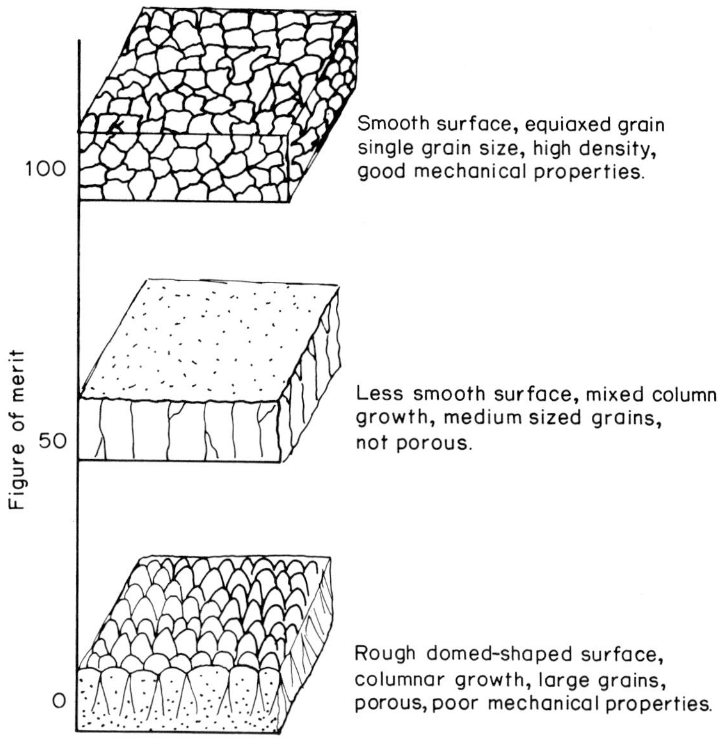

Figure 84. Figure of merit

abrasion especially on steel tools, under conditions of seizure as well as under sliding conditions (Trent, 1979). Adhesive wear may exist in metal forming where the fluid film lubricant is not sufficient to prevent adhesion between the tool and the workpiece (Uetz and Fohl, 1979). If adhesion occurs, the product is scratched and the deformation process becomes unstable (Trent, 1979). Corrosive wear occurs when sliding contact takes place in a corrosive environment. The rubbing causes the removal of the corrosion products which are poorly adherent to the surfaces and allows the corrosion procedure to repeat itself (Halling, 1979). Surface fatigue occurs when two surfaces are in rubbing contact with no abrasive or adhesive wear. That is, when the asperities slide over one another leaving one or both asperities plastically deformed (Archard and Hirst, 1956). After a critical number of contacts an asperity would fail due to fatigue forming a wear particle (Archard and Hirst, 1956).

Wear can be eliminated in different ways. It has been reported that in order to eliminate abrasive wear, the hardness of the component can be increased to exceed the hardness of the abrasive debris (Uetz and Fohl, 1979). However, the ductility of materials usually decreases with increasing the hardness thus affecting the wear resistance. If a thin, hard and well adhered ceramic coating is applied on ductile components such as steel tools, the ductility of the surface can be retained in spite of high hardness and the wear resistance is increased consequently (Zum Gahr, 1979). On the other hand, in order to eliminate adhesive wear, thin, soft metallic film lubricants can be applied to reduce the adhesion between the two sliding components (Hancock, 1968). In all cases the adhesion of the film to the substrate and the physical resistance of the components against elastic and plastic deformations have to be good in order to increase the lifetime and the performance of the coated parts. In this respect, ion plated films have exceptionally strong adherence which is generally related to the graded interface between film and substrate regardless of their compatibility (see Section 2.8a). For this reason, ion plating has been used extensively to deposit soft and hard coatings for lubrication and wear resistance. Bukley and Spalvins (1977) have studied the wear properties of ion plated germanium and silicon on bearing steels in a controlled nitrogen atmosphere using the pin-on disc test. They reported that both films reduced wear in dry sliding or when lubricated with mineral oil. Ion plated films have also been applied for the prevention of fretting (Ohmae *et al.*, 1974a, b; Ohmae, 1976) and it has been demonstrated that ion plated boron nitride films have the best resistance to fretting. Ion plated soft films have been reported to have longer durability when compared to vacuum evaporated films (Patten, 1978). The failure mechanism in this case was related to peeling off and detachment for vacuum evaporated films and ploughing and rubbing off for ion plated films. Later, it has been shown (Sherbiney and Halling, 1977) that two basic mechanisms dominate the wear of ion plated films: micro-cutting and fatigue wear. It is also claimed that the first of these is responsible for the high initial wear rates, while the second is responsible for the subsequent modest wear rates (Sherbiney and Halling, 1977). Salem (1977) studied the wear of ion plated tin and aluminium on titanium substrates in normal and corrosive atmospheres and found that the wear rates of the film–substrate interfaces are much lower than the wear rates of both the film and the substrate. It has also been found that the wear rates in the corrosive media are higher than that in normal atmosphere (Salem, 1977). Wisander (1971) tested the durability of ion plated soft metallic films and concluded that the strong adhesion of ion plated coatings was responsible for galling prevention.

In general, when thin coatings are used to improve the wear resistance of a substrate, their minimum thickness has to be a few times higher than the surface roughness of the components, otherwise the wear will proceed quickly to the base material leading to its failure under operating conditions.

6.5 FRICTION

According to the adhesion theory of friction, the friction force F between two sliding, rotating or rubbing surfaces arises from two main causes (Bowden and Tabor, 1964a, b). The first is the shearing force (S) of minute welds or metallic junctions formed between the asperities of two surfaces in contact. The second is the force (P) necessary to plough the surface asperities of the harder metal through the softer one. The total friction force is:

$$F = S + P \tag{6.5.1}$$

For surfaces which are extremely well lubricated, the shearing term, S, may become so small so that P is of a comparable magnitude. For most situations with metal surfaces, however, P is generally small compared with S so that it can be neglected. Thus:

$$F = S = S_0\, A \tag{6.5.2}$$

where S_0 is the shear strength of the metallic junctions and A is the sum of their individual areas.

Since it is these areas that actually carry the normal load W between the two surfaces, it results that:

$$W = P_0\, A \tag{6.5.3}$$

where P_0 is the yield pressure of the softer metal. By combining equations (6.5.1 and 6.5.2) an expression for the coefficient of friction (μ) can be written as follows:

$$\mu = \frac{F}{W} = \frac{S_0}{P_0} \tag{6.5.4}$$

This ratio, which refers to the softer of the two contacting metals, is fairly constant for most metals. Therefore, this simple theory indicates that the coefficient of friction of a large range of metals varies a little. However, from this simple theory it is apparent that the frictional force F is determined by the shear strength S_0 and the real area of contact A. In order to obtain low friction between the sliding surfaces, A or S_0, or both, should be small. This means that the tribological materials of high hardness and low shear strength are most suitable as lubricants. However, this is not possible in practice since metals of high hardness always have high shear strengths (Hancock, 1968). One way of obtaining lower coefficients of friction is by depositing a thin layer of soft metal of a low shear strength onto a hard metal substrate (Hancock, 1968). In this case shearing will occur in the soft metal layer while the load carrying capacity will be due to the substrate. The thin metallic film should adhere well to the substrate in order to control the friction (Ohmae, 1976). However, the extent to which the coefficient of friction can be lowered depends primarily on film thickness, surface roughness, degree of localized deformation, mechanical properties of the film relative to those of its substrate and the deposition technique (Hancock, 1968).

Ion plated thin metallic films such as gold, lead, silver and tin have been shown to offer great improvements when applied to sliding or rotating mechanical components (Holland, 1978; Spalvins, 1974). For such applications, the ion plated films, which have excellent adhesion and a graded interface, reduce the coefficient of friction and wear, and extend endurance life. Furthermore, the films alter the mode of debris generation and reduce torque noise (Ohmae, 1976; Todd and Bentall, 1978). The coefficient of friction is, however, greatly influenced by the film thickness as originally shown by Bowden and Tabor (1964) for vapour deposited indium film on tool steel. Bowden and Tabor illustrated the type of relation between the coefficient of friction and film thickness by a 'V' shaped curve where a minimum friction was achieved with a film approximately 1000–10 000 Å thick (Bowden and Tabor, 1964). For ion plated gold and lead films, Spalvins (1980) reported a similarly shaped relation where an effective thickness of 200–35 000 Å for gold films displayed a coefficient of friction of 0.1, whereas an effective thickness of 1800–2500 Å for lead films displayed a coefficient of friction of 0.085 (see Figure 85). The 'V' shaped curve is usually divided into ultra-thin and thin film regions (see Figure 86) with an effective or critical film thickness at which the coefficient of friction reaches

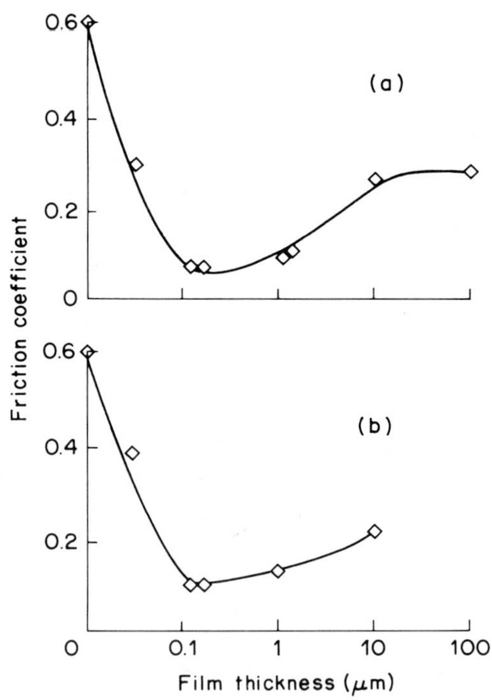

Figure 85. The dependence of the friction coefficient on film thickness for ion plated lead (a) and gold (b) films (from Spalvins, 1981) (*Reproduced by permission of Elsevier Sequoia SA*)

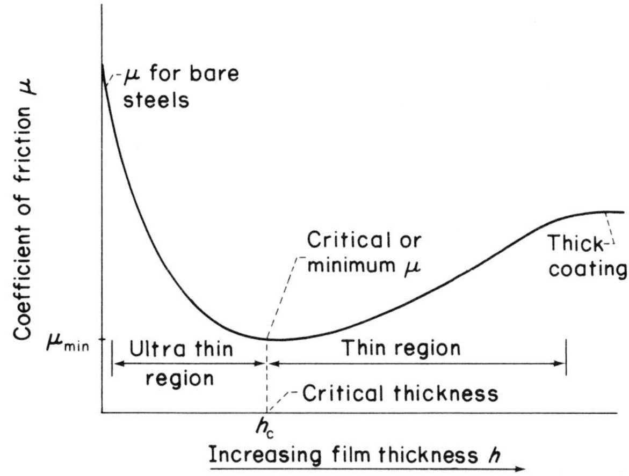

Figure 86. Schematic representation of the effect of film thickness on the friction coefficient of low shear strength metallic films on steel (from Spalvins, 1981). (*Reproduced by permission of Elsevier Sequoia SA*)

a minimum value (Hancock, 1968) The increase in the coefficient of friction in the ultra-thin region is associated with non-uniformity of the thin coatings or asperity penetration of the underlying substrate (Hancock, 1968). However, Finklin (1971) explained the behaviour of this regime by the stiffening of the film due to couple stresses. In the thin film region, as the film thickness increases, the load carrying capacity of the substrate surface decreases and an increase in the apparent area of contact leads to a higher coefficient of friction. However, the relationship between the critical film thickness and the minimum coefficient of friction is affected by the surface roughness and the deposition process. With rough surfaces the coefficient of friction increases continuously as the film thickness increases without going through a minimum (Sherbiney and Halling, 1977). Other factors which could also affect the coefficient of friction are the physiochemical phenomena such as oxidation and diffusion under certain sliding conditions (Burton, 1965; Arnell and Soliman, 1978). For example, Thornton and Wolanski (1967) investigated the frictional behaviour of gold–molybdenum composite films at elevated temperatures in atmosphere. They showed that the oxidation of molybdenum contributed to the solid lubrication at high temperatures. In a further study, Miyakawa (1978) investigated the solid lubrication of ion plated gold and silver films onto tool steel and stainless steel in vacuum at elevated temperatures. They suggested that the diffusion of some material from the substrate reduced the frictional properties of thin films. A similar study was carried out by Enomoto *et al.* (1979) and showed that ion plated molybdenum film sandwiched between ion plated gold films and a stainless steel substrate provides an effective barrier against diffusion up to 600°C in vacuum and with improved frictional properties.

As a general conclusion, ion plated soft metallic films have a lower coefficient of friction at a lower film thickness in comparison with films prepared by other deposition methods. The lower coefficient of friction is attributed to the small crystallite size, the crystallite shape and the increased cohesive strength which are the characteristics of ion plated films. It has also been observed that ion plated metallic films generally show a more gradual increase in the coefficient of friction once the film has worn through. As a result, catastrophic or abrupt failures are not observed with ion plated films. This characteristic can be explained in terms of excellent adhesion and the graded interface which prevents an abrupt seizure (Hancock, 1968).

6.6 FATIGUE

Fatigue failure is a phenomenon where components fail at stress levels below their design strength and crack initiation sites are difficult to detect before failure. Ion plating has been shown to extend the fatigue life of components by many times as compared to other types of coating. Spalvins (1980) reported that nickel and inconel fatigue specimens plated with a 1500 Å thick gold and copper film had shown 23% to 27% increase in fatigue life respectively (Figure 87) and that all fatigue failures started at the surface. Similar results have also been achieved

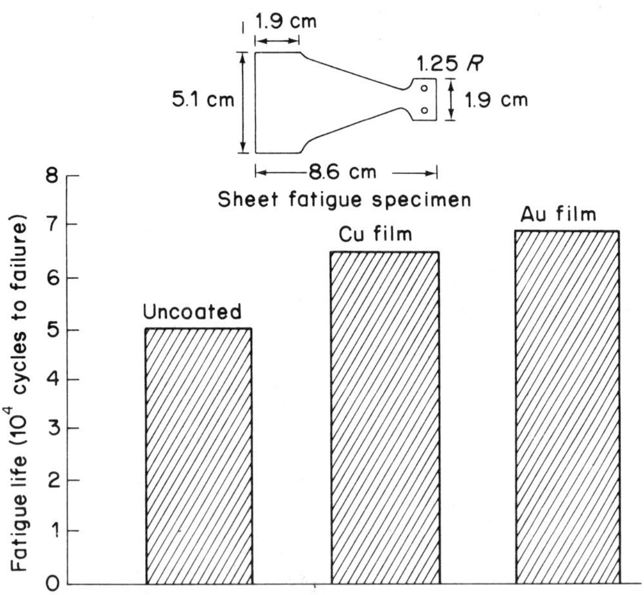

Figure 87. Effect of copper and gold ion plated films (1500 Å) on the fatigue life of nickel (from Spalvins, 1980). (*Reproduced by permission of the American Institute of Physics*)

for ion plated gold on steel. Work carried out at Osaka University (Ohmae *et al.*, 1974; Ohmae 1976) on the fatigue properties of low carbon steel has shown an improvement in the fatigue limit of ion plated gold on steel in laboratory air, whilst electroplated films have no noticeable effect. Such an improvement was related to the dense structure of ion plated films which contain neither pit nor pore and to the graded interface where solid solution strengthening and reduction of surface residual stresses occurs during ion plating (Nakai *et al.*, 1974).

Pits, pores or even spits (see Section 3.3(a)) are known to be stress raisers and sites for fatigue crack initiation although they do not affect the yield strength or reduction of area in a ductile material (Boone *et al.*, 1974). Ohmae *et al.* (1974) have further shown that annealing the ion plated films resulted in a great increase of fatigue limit and this effect was related to the thermal diffusion of gold into the steel surface. Furthermore, in a recent comprehensive study, the Spring Research and Manufacturers Association (SRAMA) reported a significant improvement in fatigue resistance of ion plated aluminium on springs by a factor of 100 as compared to springs coated by other methods, including electroplated zinc and cadmium, even after 96 hours exposure to neutral salt spray corrosion testing (Reynolds, 1985). Ion plated aluminium has also been found to increase the fatigue life of aircraft fastener holes (Morris, 1976). Figure 88 shows the surface finish of holes versus the fatigue life for 2024-T aluminium. As demonstrated by this graph, sulphuric acid anodized coatings reduce the fatigue life of holes in which Taber-Lock interference fasteners were installed by 50%. Peening prior to anodizing only results in a 25% recovery of the 50%

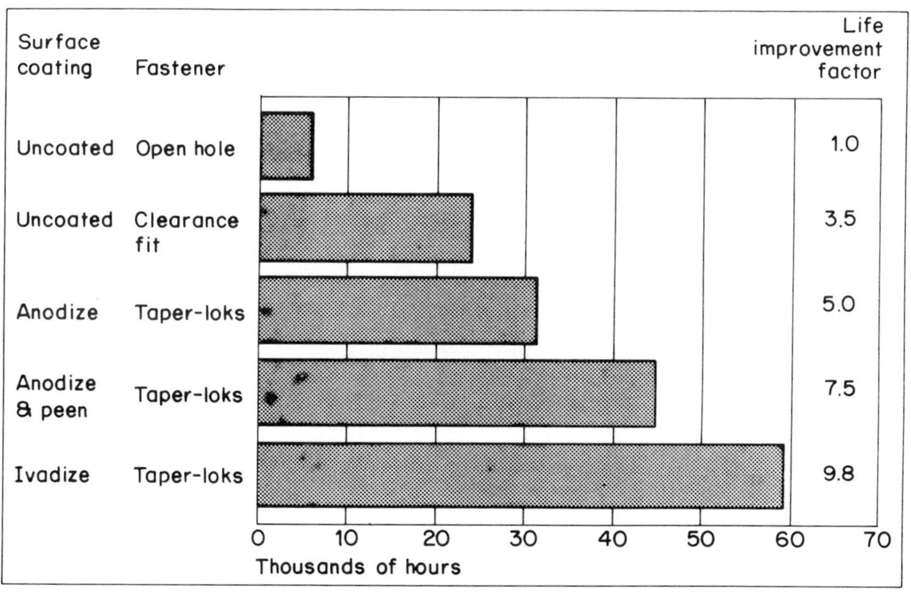

Figure 88. Surface finish of holes versus the fatigue life of 2024-T aluminium (*Reproduced by permission of McDonnell Douglas Corp.*)

loss in fatigue life. Coating the test piece with ion plated aluminium has no appreciable adverse effect on fatigue life and is lower in cost to apply than peening. For this reason, aircraft parts such as wings are currently ion plated with aluminium using the commercial process known as Ivadizer (Section 7.1) for corrosion protection and to improve the fatigue properties (Ahmed, 1986).

Chapter 7
Applications of the Ion Plating Technique

The applications of the ion plating technique have been increasing steadily and it has now found use in widely differing industrial purposes, ranging from production of films for decorative uses to corrosion protection. However, since each application has required its own set of processing variables and designs, a wide range of new equipment design has evolved. This, in turn, has led to even more new applications. Table 4 summarizes the typical applications of ion plating. The process is now used commercially, particularly for the coating of aircraft fasteners and other parts such as wings and for titanium nitriding. These applications with others will be reviewed in detail in the following sections.

7.1 CORROSION PROTECTION

Cadmium electro-plated steel components and fasteners have been used in aircraft structures and in other industrial products for many years (Fager and Spuir, 1970, 1971). The plating cathodically protects the steel and prevents accelerated galvanic corrosion at contacts with aluminium alloys. Zinc has also been widely used as a sacrificial coating for steel, but cadmium was preferred in aircraft structure for many reasons. Cadmium is more galvanically compatible with aluminium alloys, cadmium plated bolts exhibit lower frictional properties and they are far less prone to seizure due to the wedging action of corrosion products (McLoughlin, 1978; White, 1978). However, in recent years concern has increased over problems caused by cadmium electro-plating such as exfoliation corrosion in the countersinks of high strength aluminium alloys secured with cadmium plated fasteners (Fager and Spuir, 1971; Fannin and Steube, 1974). Furthermore, on high strength steel, cadmium electro-plating can cause hydrogen embrittlement. Cadmium has also been found to cause failures of titanium fasteners at relatively low temperatures (Taylor, 1969). Cadmium is also toxic to human beings and harmful to the environment (Brande, 1977) and its use is expected to be banned in the near future.

For corrosion protection there are two general methods by which coating can protect a substrate (Nowak *et al.*, 1985):

(1) The coating is electro-chemically more active than the substrate in the environment.
(2) The coating is less active (more noble).

In the first method, the coating is sacrificial and it usually gives substantial protection even when porous. In the second method, the coating must be

Table 4. Applications of Ion Plating Technique.

Application	Materials of Coating	Materials of Substrate	Thickness (micron)	Deposition Rate	Production Details	Substrate Temperature
Corrosion protection	Al	U, Steel, Titanium alloy	Up to 40	2 micron/min	100–150 lb of fasteners per hour	250–300°C
High temperature corrosion protection	MCrAlY	Ni or Co base super alloy	125	5 micron/min		800–1100°C
	Cr	Steel				
	Al_2O_3	Steel				
	Si_3N_4	Mo				
	Cd	Steel				
Wear resistance coating	TiN	Cemented carbide, high speed steel, die steel	1–5	5–10 micron/hour	Up to 1400, ¼″ dia round shank tools per 2.5 hour	400–500°C
	TiC					
Electrical	Pt, Al, Au, Ag	Si	1			
	Al	GaAs				
	In (Ga)	CdS				
	Cu	Al_2O_3				
	Si_3N_4	Si–Mo		0.7 micron/min		250–300°C

Lubrication	Pb, Au, Mg, MoS_2, Al	Metals	~1000 Å – 1 micron	
	PbSn alloy, graphite	Plastic		
Optical and Decorative	Cr	Plastic		50–100°C
	In(Sn)O	Glass, plastic	Up to 10 micron	
	TiN	metals		
	ZrC			
Bonding	Ag	Be	Up to 10 micron	
	Cu	Ta, W, Nb, oxides		

non-porous and robust, otherwise the corrosive medium will find its way through pores and attack the substrate. Programmes aimed at finding a suitable material and a process to replace cadmium electroplating have concluded that aluminium is the most promising alternative (Fannin, 1977; Fishter, 1978; Muehlberger, 1978) and the ion plating technique is the best in providing dense, adherent coating.

In 1974, the McDonnell Douglas Corporation developed an ion vapour deposition (IVD) machine called 'Ivadizer' for applying aluminium coatings on aircraft fasteners and for plating large articles. The principle mechanisms of the IVD are similar to ion plating where the substrate is biased with negative voltages and the aluminium is evaporated in an argon glow discharge, with the exception that high evaporation rate and hence low ionization coefficient is used in the IVD process. Figure 89 shows a schematic diagram of two IVD industrial machines, namely a barrel coater to deposit aluminium on fasteners and a rack coater to deposit aluminium on large items. Figure 90 shows pictures of the standard barrel and rack coater.

The application of ion plated aluminium for corrosion protection is now well documented and recognized by a number of government, industrial and commercial specifications, particularly in the United States and Japan (MIL-C, 1976). These specifications divide the aluminium coating thickness into three classes and two types for different corrosion protection applications. Such classes are commercially produced in the Ivadizer unit at 1 kV bias and 1 Pa argon pressure and are shown in Table 5. Type I is ion plated aluminium and type II is ion plated aluminium with a chromated top coat. The supplemental chromate treatment is applied for additional corrosion protection and for the adhesion of subsequent painting applications. Figure 91 shows a typical type II coating on fasteners. The following reviews some specific applications of ion plated aluminium coatings on various substrates for corrosion protection.

Table 5.

		Corrosion Test Period (hours)	
Class	Thickness	Type I	Type II
1	25 micron (minimum)	504	672
2	12.5 micron (minimum)	336	504
3	7.5 micron (minimum)	168	336

Class 1 coatings are used in severe corrosion environments.
Class 2 coatings are used in less severe interior corrosion environments.
Class 3 coatings are used when fine tolerances are necessary on such parts as fine threaded parts
(*Reproduced by permission of McDonnell Douglas Corporation, 1986*)

7.1(a) The Corrosion Protection of Steel and Titanium

Ion plated aluminium coating on steel and titanium usually offer good corrosion protection after exposing to both 5% neutral salt fog and SO_2–salt spray environments (Ahmed, 1985a). Figure 92 shows a comparison between the

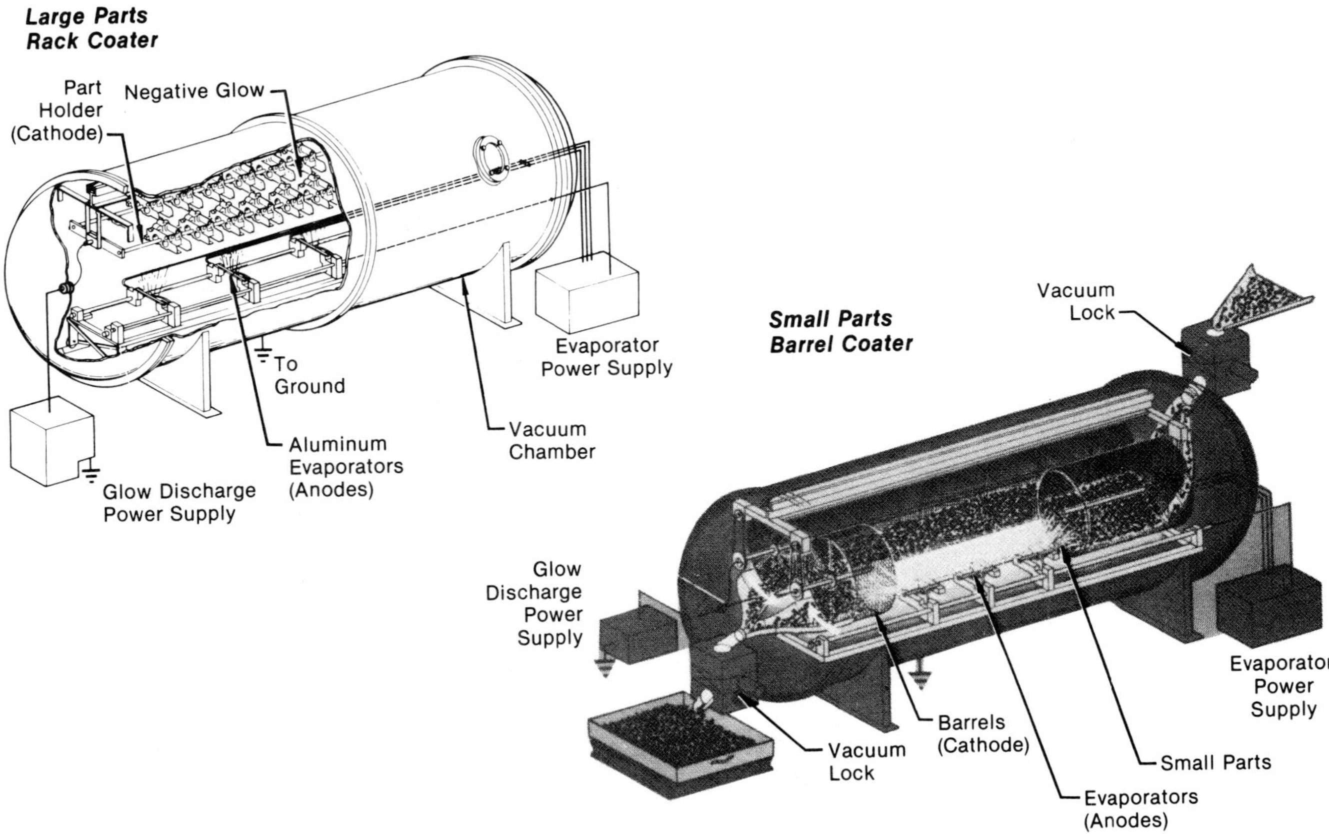

Figure 89. Schematic diagram of barrel and rack Ivadizers. (*Reproduced by permission of McDonnell Douglas Corp.*)

(a)

(b)

Figure 90. Standard (a) barrel and (b) rack Ivadizers. (*Reproduced by permission of McDonnell Douglas Corp.*)

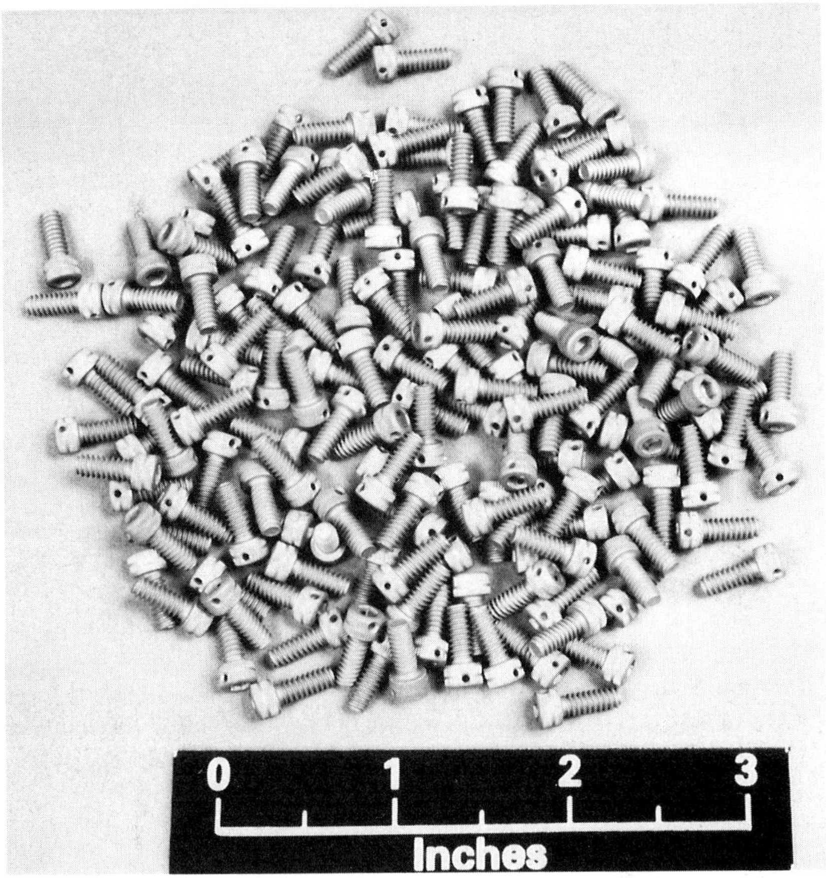

Figure 91. Typical type II coating. (*Reproduced by permission of McDonnell Douglas Corp.*)

corrosion protection of cadmium and ion plated aluminium after 168 hours in SO_2–salt fog test. The aluminium is cathodically protecting the steel substrate and, therefore, suffers less corrosion. Table 5 shows the effect of different thicknesses of ion plated aluminium on the corrosion protection of the steel substrate. Figure 93 shows a comparison between ion plated aluminium on titanium fasteners and uncoated titanium fasteners after 24 days testing in SO_2–salt spray. The test shows the appearance of blistering around the peripheries of the bare titanium fasteners which is much more severe than around the ion plated fasteners. Although cadmium generally gives better protection of steel than ion plated aluminium, should a scratch be made through the coatings, cadmium tends to sacrifice itself more quickly and allows red rust to appear before the ion plated aluminium does. Furthermore, cadmium can embrittle high strength steels (Fager and Spuir, 1970) and titanium alloys

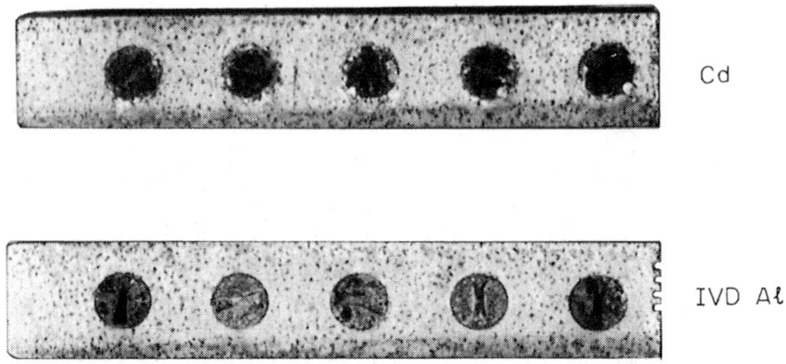

Figure 92. Comparison between corrosion protection of cadmium and ion plated aluminium on steel after 168 hours in the SO_2 salt spray test. (*Reproduced by permission of McDonnell Douglas Corp.*)

(Muehlberger, 1978). It has also been shown recently that steel springs ion plated with aluminium showed good corrosion protection and fatigue performance as compared to other coatings including zinc and cadmium electro-plating (Reynolds, 1985). The test showed that the fatigue resistance of ion plated springs is 100 times more than springs coated by electro-plating.

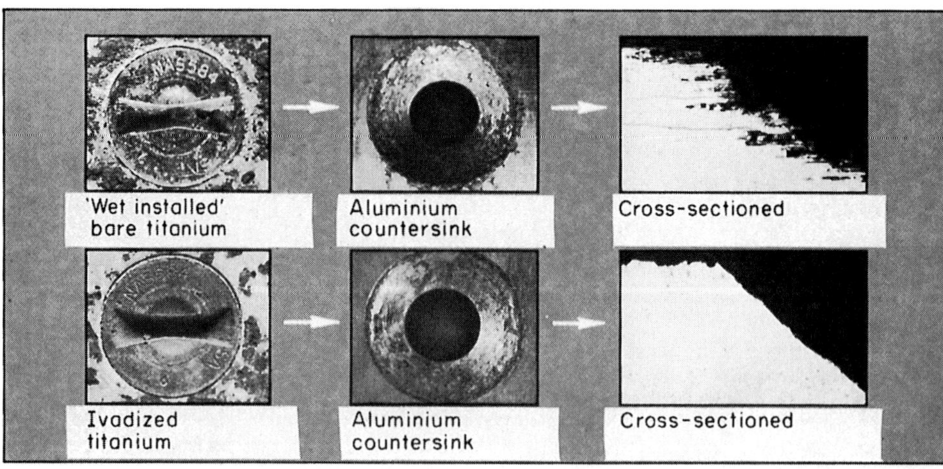

Figure 93. Ion plated aluminium on titanium and bare titanium after 24 days in the SO_2 salt spray test. (*Reproduced by permission of McDonnell Douglas Corp.*)

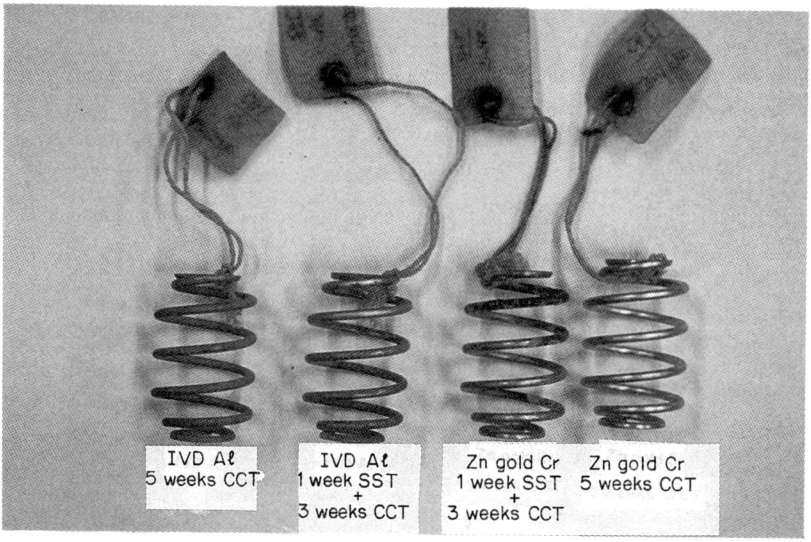

Figure 94. Ion plated aluminium and zinc gold chromated on springs after 1 week in salt spray (SST) followed by 3 weeks in a condensation chamber (CCT) and for 5 weeks in a condensation chamber only (from Ahmed, 1986a). (*Reproduced by permission of The Institute of Metals*)

7.1(b) The Corrosion Protection of Copper Alloy

The corrosion protection of ion plated aluminium coating on copper alloy springs was compared to similar springs which were electro-plated with zinc–chromate (Ahmed, 1985a). The springs have been tested in a salt spray chamber (ASTM B-117) and a condensation chamber (DIN 50017). Table 6 summarizes the results of the corrosion test. Figure 94 shows the two types of coating after 1 week in a salt spray test followed by 3 weeks in a condensation chamber test and after 5 weeks of continuous testing in the condensation chamber. Figure 95 shows a comparison between ion plated aluminium coated and zinc chromate springs after 1 week in a salt spray test followed by 3 weeks in a condensation chamber. It is well evident that ion plated aluminium provides good corrosion protection to copper alloy springs.

7.1(c) The Corrosion Protection of Aluminium Alloy Fuel and Pneumatic Line Fittings

Ion plated aluminium coating can provide the necessary electrical bond across the fitting in order to dissipate static electrical charges generated by fuel or airflow and also to protect the fitting against corrosion (Figure 96).

Table 6 Subjects: Ion Plated Al on Springs, Class 2 (13 micron), MiL-C-83; 488 ZnCr-coated Springs, 5 micron Zn-chromated.

	Salt spray test (hours SST)	4h	28h	48h	72h	120h	168h	+ 3 weeks CHT
Salt spray chamber (Redrust/Whiterust) ASTM B117	Al	10/10	9/10	9/10	9/9	9/9	9/8–9	8/5 (*)
	Zn	10/10	10/9	10/8	10/7	10/7	10/6	4/4
	Condensing Humidity Test (CHT) (hours)	4h	28h	48h	120h	168h	792h	—
Condensing Chamber (Redrust/Whiterust) ASTM B117	Al	10/10	10/10	10/10	9/9	9/9	9/7	—
	Zn	10/10	10/10	10/10	10/10	10/9	9/8(*)	—

*Pitting/Pores

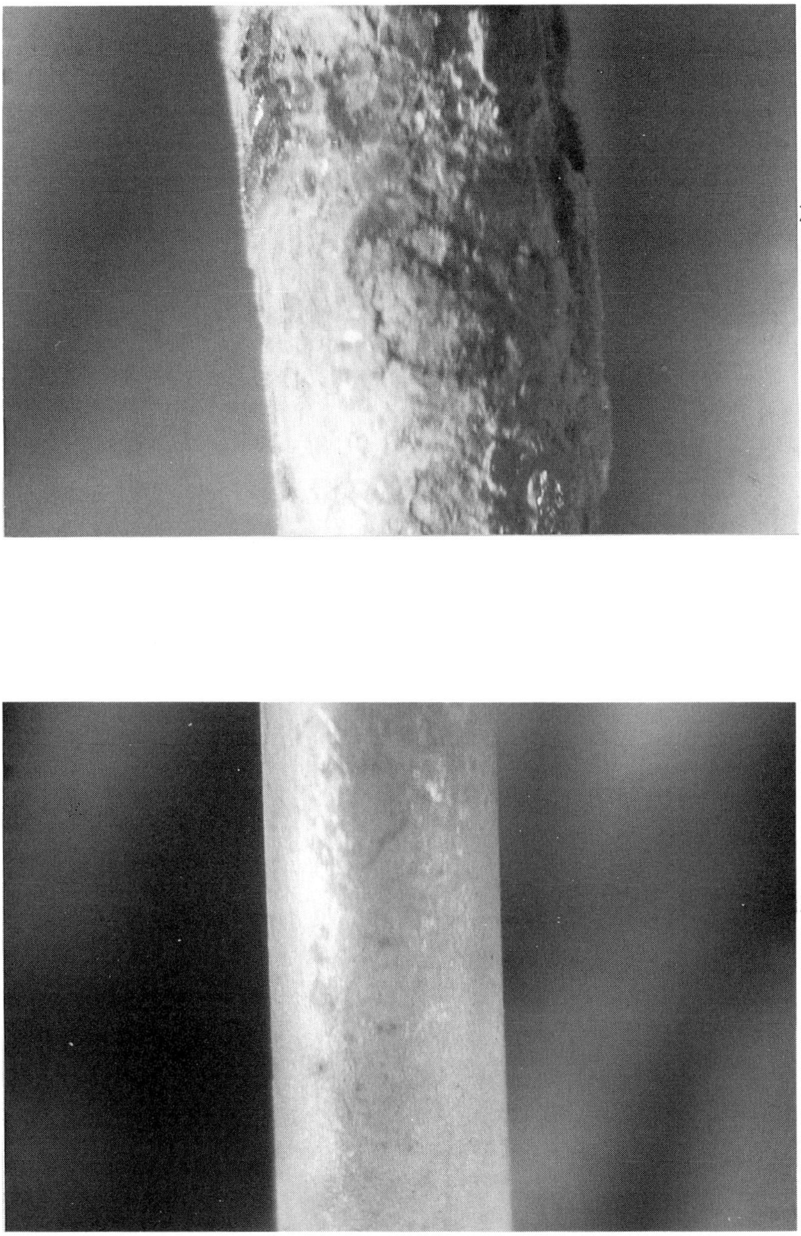

Figure 95. Ion plated aluminium and zinc gold chromated coatings on springs after 1 week in salt spray and 1 week in a condensation chamber (from Ahmed, 1986a). (*Reproduced by permission of The Institute of Metals*)

Figure 96. Ion plated aluminium on fuel and pneumatic line fittings. (*Reproduced by permission of McDonnell Douglas Corp.*)

7.1(d) Aluminium Alloy Electrical Connectors

Ion plated aluminium has a distinct advantage over cadmium on aluminium alloy connectors. Cadmium does not adhere well on aluminium alloy and a copper or nickel has to be used as a binding layer. Ion plated aluminium adheres well to aluminium alloy and provides the required electrical conductivity along with corrosion protection to increase salt spray requirements to 1000 hours (Figure 97). It can also replace nickel plating for high temperature applications.

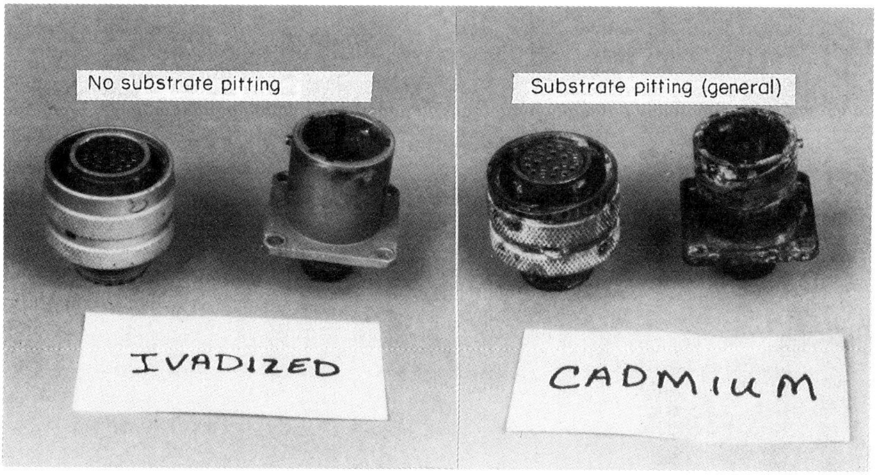

Figure 97. Ion plated aluminium on electrical connectors. (*Reproduced by permission of McDonnell Douglas Corp.*)

7.1(e) High Temperature Applications

Ion plated aluminium coating has a useful service temperature of 350°C. This feature and its excellent corrosion resistance has led to its applications on the aft engine hangers of the DC–10 and Boeing 767 aircraft. United Airlines reported that before the use of ion plated aluminium it was necessary to remove and refurbish DC–10 aft engine mounts every 1500 to 3000 flight hours. They reported that some engine mounts which were coated with ion plated aluminium (Figure 98) have endured 9000 flight hours without deleterious effects (Ahmed, 1986a).

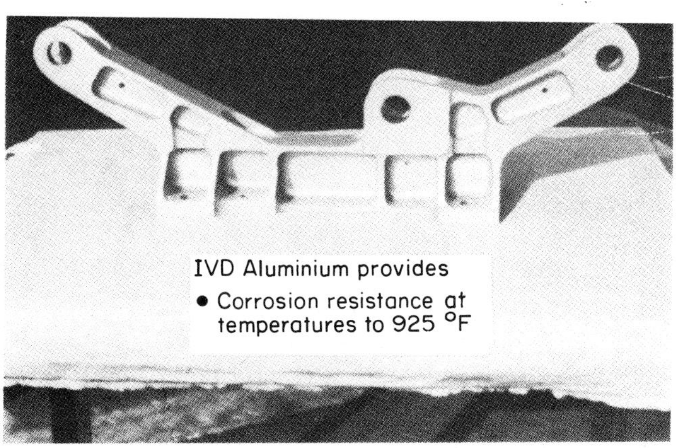

Figure 98. Ion plated aluminium on aircraft hanger. (*Reproduced by permission of McDonnell Douglas Corp.*)

7.1(f) Depleted Uranium Substrates

Aluminium coating of uranium reactor parts was one of the first applications of ion plating (Bland, 1968). Investigation of this process was based on the knowledge that aluminium provides a low neutron cross-section, has high thermal conductivity and has electro-negativity close to that of uranium so that galvanic corrosion is minimized. The reactor elements were first sputter cleaned followed by ion plating of aluminium. It was further shown that subsequent heating of the deposited layer to 600°C resulted in a mixture of UAl_2 and UAl_3 which provides a thermodynamically stable barrier to any reactor corrosion (Bland, 1968).

US Army tests have also shown that ion plated aluminium offers the best protection against corrosion and stress corrosion cracking on depleted uranium penetrator cores (Muehlberger and Reilly, 1983). Ion plated aluminium will be used pending successful field tests.

7.2 WEAR RESISTANCE CERAMIC COATINGS

In recent years, the development of processes for depositing hard, wear resistance coatings for different industrial applications has become increasingly important (*Ion Plating and Allied Techniques*, 1977–1986). Of the many processes utilized to deposit ceramic coatings such as chemical and physical vapour deposition techniques, the reactive ion plating process has been found to be suitable for such a purpose. One of the distinctive advantages of the ion plating process is its ability to deposit wear resistance ceramic coatings on high speed steel tools at temperatures below the tempering temperature of the steel type and also below temperatures at which dimensional changes can occur (Matthews, 1985). The stimulus for such development was the widely held conjecture that the advantage of wear resistance coatings on parts and tools made from steel would be, at least, as great as cemented carbide tips (Hatschek, 1983). Therefore coatings such as TiN, TiC, Ti(CN), Cr_3, C_2, Al_2O_3 and their combinations are frequently applied on twist drills, cutting tools, gear cutters, dies and moulds, electrical contacts, watch cases and bracelets (Hintermann, 1984). However, one of the most successful ceramic coatings currently used in industrial applications is titanium nitride (TiN), because of its high hardness, thermal stability and corrosion and erosion resistance (see Section 5.2(a.1)). In addition, because of the rapidly increasing price of gold it has also become desirable to coat articles with a shiny gold-like colour and TiN, having such a lustre, is an excellent substitute. Due to the importance of titanium nitride coating in many industrial applications the subject has become the main theme of many publications and conferences. Today, there are many companies around the world offering ion plating industrial machines for titanium nitride coatings for different applications, ranging from units for coating twist drills and cutting tools to units for coating articles with decorative yellow gold finish (Coating Companies). Titanium nitride coating deposited by such industrial machines

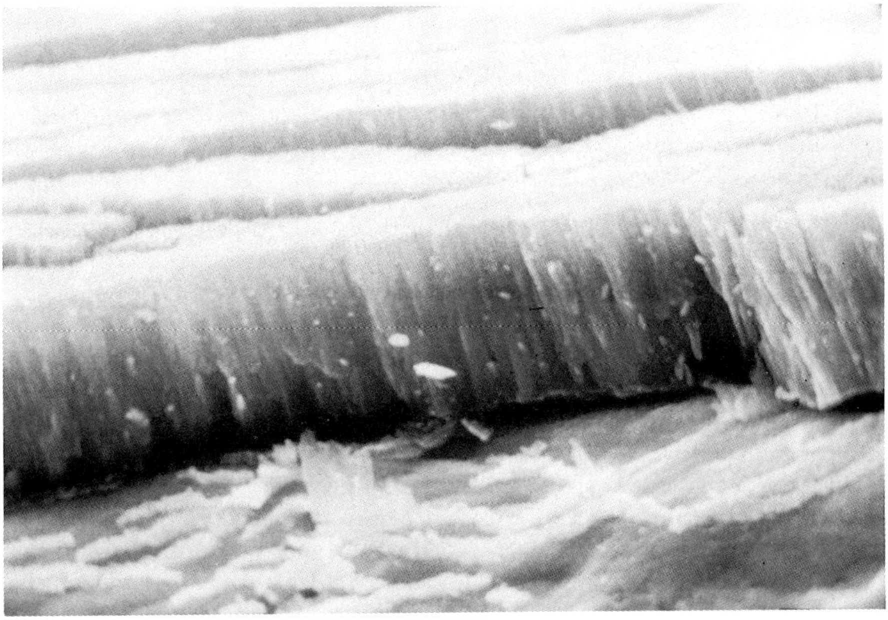

Figure 99. Structure of ion plated TiN on steel

has, in general, a smooth and dense structure like that shown in Figure 99, even at low substrate temperatures (Larden *et al.*, 1978). Therefore, polished substrates will retain their mirror surface after coating and no subsequent polishing is necessary. Furthermore, the dense structure is important in practice since a coating composed of weakly bonded columns can never achieve the full hardness of the bulk material. In addition, the adhesion of TiN to the base material and the intrinsic cohesion of the coating play an important role in increasing the tool life and the performance of the work pieces (Bunshah and Schramm, 1977; Buhl *et al.*, 1981). This is a critical requirement since the locally high shear stresses applied to the coating during operation have to be transmitted to the substrate. This subject will be discussed in detail in the following sections.

7.2(a) Applications of TiN Coatings

7.2(a.1) Cutting Tools

One of the major applications of ion plated TiN is in cutting tools, particularly high speed steel (HSS) twist drills. As compared with carbides, high speed steels have greater resistance to shock, even when fully hardened, are readily shaped by forging and machining and are available in greater number of forms. They also retain enough hardness to cut metals at reasonable rates even at high

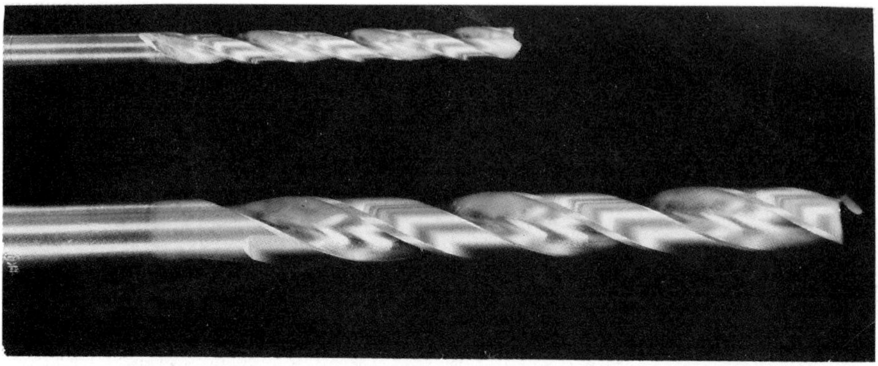

Figure 100. Typical twist drills coated with TiN using reactive ion plating

temperatures (up to 900°C) with good wear resistance and low basic cost. For these reasons, high speed steels are the preferred cutting tool material for a variety of metal cutting operations (Sherbiney and Aboukhashaba, 1983). Interests in extending the life of HSS twist drills by TiN coating have shown a number of advantages. First, the wear of the drills by hard and abrasive materials, is reduced, as is the tendency to form a built-up edge. This means that a TiN coated drill is able to make four times as many holes in a high strength steel as an otherwise identical uncoated drill (at the same cutting speed and feed rate) before regrinding becomes necessary. Figure 100 shows a typical twist drill

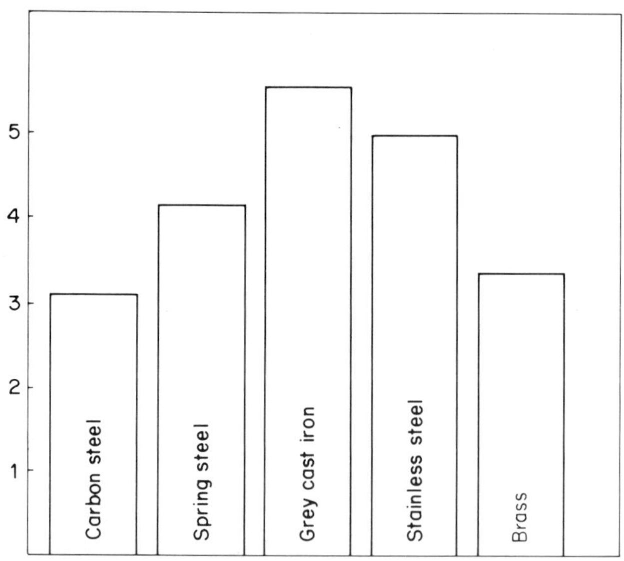

Figure 101. Improvement in twist drills' lifetime after coating with TiN (from Buhl *et al.*, 1981) (*Reproduced by permission of Elsevier Sequoia SA*)

coated with TiN using reactive ion plating. The improvement in the lifetime is shown in Figure 101 for several materials drilled (Buhl *et al.*, 1981). The improvement in metal cutting by the coated twist drill is also evident in the nearly 20% reduction in the feed power required at a given feed rate. An additional benefit is that the holes produced after have a much better surface finish than those made by an uncoated drill because the cutting is much more regular. It has also been reported that coated drills still provide an extended life even after being reground, (Figure 102) (Konig, 1982). TiN coatings can reduce the machining time by operating at higher speeds, or maximize the tool life by machining at conventional rates.

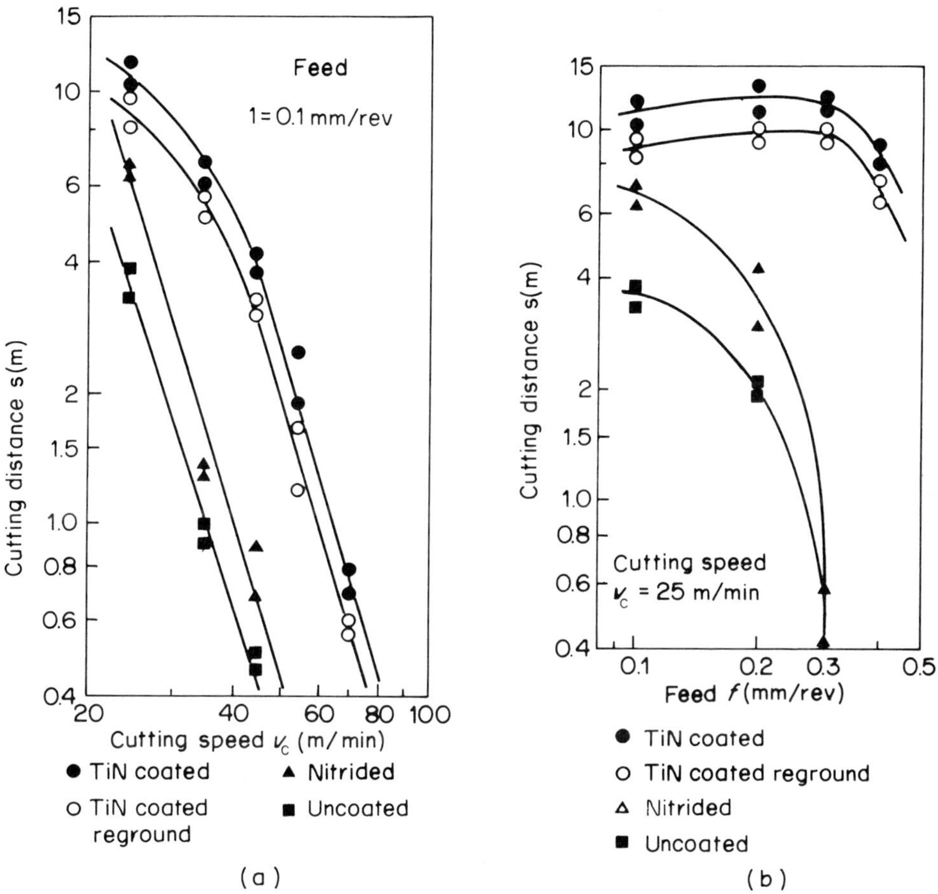

Figure 102. (a) Dependence of the cutting distance on the cutting speed at constant feed of drills with and without surface treatment (*Reproduced by permission of Prof. H. Konig, from Industrie anzeiger, Nr 53 V.2.7, 1982/104 Jg.*) (b) Dependence of the cutting distance on the feed rate at constant cutting speed of drills with and without surface treatments. (*Reproduced by permission of Prof. H. Konig, from Industrie anzeiger, Nr 53 V.2.7, 1982/104 Jg*)

There are now many published test results for coated drills, with several manufacturers offering their own products. In an extensive study Sproul (1984, 1985, 1986) showed that TiN coated drills gave an average life of 5306 holes 35 mm in diameter in 4340 steel. However, the standard deviation was 407 holes. Studies have shown that drills are unique in producing this level of scatter in performance, probably partly due to variability in the base material, its hardness, surface finish and geometry. Sproul (1986) has also found that the type of cutting fluid used with these coatings could markedly affect tool life. For example, if one fluid doubles the number of holes drilled, another fluid with a high sulphur content reduces the average number of holes cut from 150, when unlubricated, down to 10. The reason for this is not clear yet and therefore users must be careful in selecting the suitable cutting fluids for use with these coatings (Sproul, 1986).

Another potential application for TiN coatings is in gear cutting hobs where the accuracy of the completed gear depends on the condition of tool during the last finishing cut and, therefore, is greatly affected by the wear of the hub (Sundquist *et al.*, 1982). Wear resistant coatings are an attractive solution to this problem. When using sintered solid steel hobs, reactive ion plating has proved to be a suitable coating process and TiN a beneficial coating material. Sunquist *et al.* (1982) studied the flank wear of coated and uncoated gear hubs in a production line. The cutting conditions were kept constant in both cases. In these tests 2 to 3 times more gears were cut between reshaping operations with coated tools than with uncoated tools (Table 7).

There are several advantages of coating gear hubs with TiN. The first and the most important advantage is the economic benefit in increasing the life of the hubs by at least 4 to 5 times. The other advantage is that the scatter in results mentioned earlier is wider for twist drills than for large cutting tools. This has been attributed to several factors including an increased risk of overheating and softening small diameter, low thermal mass items during coating (Matthews, 1984).

7.2(a.2) Forming Tools

Applications of TiN coatings have also been investigated in the metal forming field, including metal bending where metal sheets are bent with a tool to produce

Table 7. Cutting Parameters and Number of Gears Produced Without Resharpening*.

Cutter (ϕ 100 mm $z=33$)	Workpiece Hardness (HB)	Cutting Conditions				No. at Gears Produced	W_f (mm)
		V_1(m/min)	V_2(m/min)	S_w(mm)	S_r(mm)		
Uncoated	311	9	12	0.5	0.03	4	0.2
Coated	311	9	12	0.5	0.03	18	0.2

Where V_1 = cutting speed, 1st cut; V_2 = cutting speed, 2nd cut; S_w = rotary feed; S_r = infeed/stroke; W_f = flank wear.
*From Sundquist *et al.* (1982). (*Reproduced by permission of The Institute of Metals*).

light hinges. The bending groove of the tool has to be ground and polished to ensure good surface finish. The only lubricant used in this process is the oil which remains on the sheet surface after rolling. By applying TiN coating on the bending tool, Sundquist *et al.* (1982) extended the life of the tool by 50 times compared to the uncoated tool. Coated tools first produced 10 000 hinges without any scratching, and after removal of adherent wear residues, the tools produced 50 000 hinges before the coating wore through. In comparison, uncoated tools had to be reground and polished after producing 1000 hinges. Therefore, the benefit of the coating is far greater than only cost saving, as a considerable amount of down time and setting-up cost can be avoided.

There has also been a growing interest in TiN coating tools for extrusion moulding and for deep drawing. TiN coated tools have been tested in competition with hard chromium coated tools and showed an increase in the number of parts made by a single tool by a factor of up to 4 times (Buhl *et al.*, 1981). Therefore, the particular advantage of the TiN coating on extrusion moulding tools is the prevention of cold welding between the tool and the parts, which otherwise lead to a roughening of the tool surfaces and its rapid failure. Other applications of TiN coating on tools include plastic moulding dies, piercing punches, paper cutting knives, nozzles for abrasive shaping (Matthews, 1984), and surgical tools (Gordashnik *et al.*, 1983). However, the applications of TiN coating are expected to extend to other fields in the near future.

7.2(a.3) Decorative Coatings

The yellow golden colour of TiN has made it very attractive in decorative applications to replace gold (Clarke, 1977; Munz *et al.*, 1983; Matsumura *et al.*, 1983). The colour of TiN can be controlled by process parameters such as nitrogen partial pressure and substrate bias. With a low partial pressure, the film exhibits a rather silver-white colour and by increasing the partial pressure, the colour changes from white gold to bluish gold then to yellow gold. Furthermore, the high abrasion resistance of TiN coating eliminates the need for an organic top coat to protect the finish. However, a spray epoxy lacquer is used sometimes on rings or jewellery with intricate artistic detail in order to provide contrast between light and dark areas. In such applications, however, the substrate, which is made of 80 per cent nickel/20 per cent chromium, is polished to a jewellery finish using buffing wheels and compound, then coated with TiN. The original polished condition of the uncoated article is retained after the TiN coating process because the surface finish is practically perfectly reproduced by the coating process. Tarnish tests involving exposure to a sulphide atmosphere for 24 hours showed no visible evidence of discolouration by tarnish products. However, when tested under similar conditions, 10 carat yellow gold developed a pale red/brown tarnish film, while 2 carat gold–silver alloys exhibited a very dark blue/black film after 6 hours of exposure (Report, 1985). In other applications such as watches, the cases are made from steel or brass which is coated with a hard corrosion-resistant load bearing layer (e.g. electrolytically

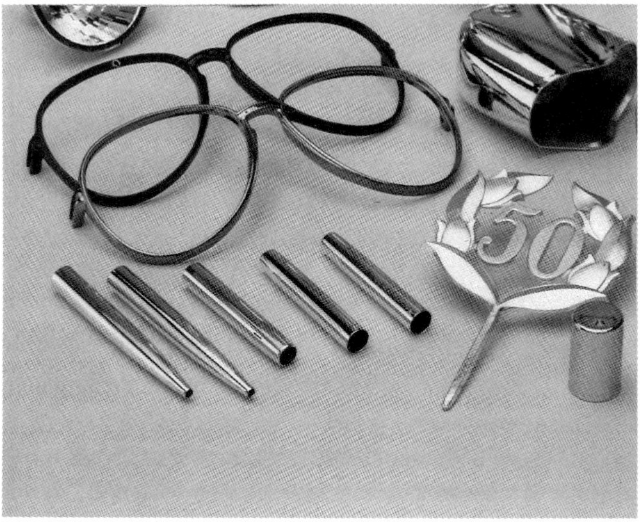

Figure 103. Decorative applications of TiN coating

deposited hard chromium) followed by the gold-coloured TiN coating. Figure 103 shows some of the current decorative applications of TiN coating.

7.2(a.4) Ball Bearings

It has been reported by many workers that the friction and wear properties of hard coatings such as TiN and TiC under oil lubricated conditions have improved bearing behaviour (Hintermann, 1980; 1981; Boving *et al.*, 1981, 1983). For lubricated sliding concentrated steel contacts the load bearing capacity can be increased considerably and in high precision/high speed ball bearings, using coated balls or rings, abrasive wear and decomposition of the oil can be greatly reduced. Further evidence of the beneficial effect of hard, wear-resistant, friction reduced coatings has been obtained from ball bearing applications under extreme conditions such as high temperature (Hintermann *et al.*, 1978).

Hintermann *et al.* (1978, 1984) summarized the advantages of coating stainless steel (AISI 440C) gyro bearings with TiC as follows:

(1) The coated bearings are stiffer than the uncoated ones.
(2) The bearings with either TiC coated balls or rings operated very satisfactorily at 24 000 r.p.m. for 25 000 hours in one case and more than 20 000 hours in others.
(3) Compared to standard uncoated bearings, no measurable cage-wear can be observed when coated balls were used.
(4) The lubricant does not deteriorate since very little debris is produced and the latter is not as catalytically active as that formed from steel parts. The oil remains transparent, uncoloured, clear.

(5) The bearings with coated balls run smoother than standard uncoated bearings.
(6) Even with low quality coated bearing parts the bearings performed much better than standard uncoated bearings under the same conditions.
(7) Basically, similar results are obtained whether the TiC coating is on the races or on the balls.

Furthermore, Hintermann *et al.* (1978) and Hanni and Hintermann (1979) have also reported that TiC and TiN have a remarkably high corrosion resistance in most corrosive media such as H_2SO_4 and that such resistance is higher than that of stainless steels and Ni–Co super alloys. However, such resistance only applies if coatings are absolutely dense and hence free from open pores and micro-cracks. For these reasons, NASA's George Marshall Space Flight Centre is planning to use ion plated hard coatings such as TiC on 440C stainless steel bearings for use in the Space Shuttle's main engine (Technology Forecast, 1986).

7.3 LUBRICATION

Long term operation of moving parts such as bearings and bearing materials usually require the application of a liquid lubricant film to the surface to reduce adhesion, friction, and wear. However, under certain conditions where fluid lubrication is beset with difficulties such as at high and low pressures, high and low temperatures and slow oscillations, or where corrosion is a problem, a dry solid lubricant coating can be used, either on its own or in conjunction with a conventional fluid lubricant (Bisson, 1964; Spalvins, 1971). Thin, low-friction films such as Al, In, Sn, Ag, Au, Pb, PbSn alloys, graphite and MoS_2 have been used on hard substrates to replace conventional lubricants. Applications of such dry lubricants include the coating of bearings and seals for vacuum or low temperature (liquid gas) environment, moving parts in radioactive areas or reactive gas atmosphere and plastic and polymers that degrade with oil lubricants (Bisson, 1964). Of the soft metallic films such as Au, Ag and Pb, gold and lead have found extensive uses in space born bearings of satellite mechanisms, including solar array drives, despin assemblies and gimbals (Spalvins, 1971). These ion plated lubricating thin films of 0.2 μm in thickness are very effective in improving the frictional and wear characteristics and increasing the endurance drives (Hancock, 1968; Hart and Cutler, 1973). Figure 104 shows typical friction curves for ion plated and vacuum deposited gold films 0.2 μm thick on 440C steel with respect to testing time in a pin-on-disc tribotester under vacuum conditions. Spalvins (1971, 1980) reported that such a film has three distinct improvements over vacuum deposition: (1) increased endurance life, (2) lower coefficient of friction and (3) avoidance of catastrophic failure. Furthermore, he attributed the increased endurance life to the good adhesion, the lower coefficient of friction due to the dense, cohesive small crystalline size and thin film thickness. The gradual increase in the coefficient of friction after the film has been worn off during use was attributed to the formation of the graded interface.

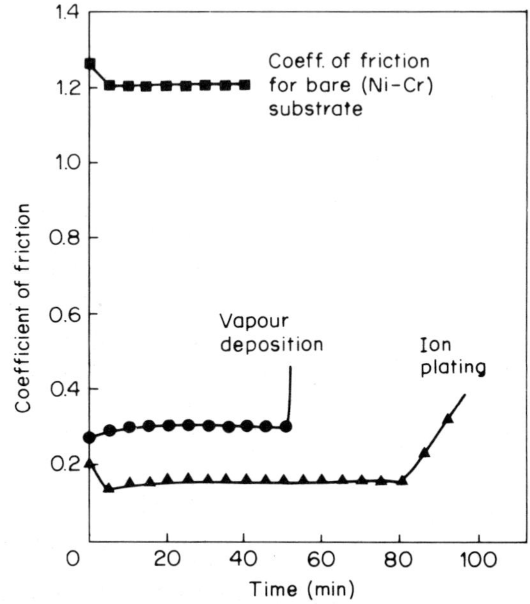

Figure 104. Coefficient of friction of niobium sliding on (Ni–Cr) alloy with gold deposited by vacuum deposition and ion plating (from Spalvins, 1980). (*Reproduced by permission of the American Institute of Physics*)

Another promising development in the field of lubrication by thin metal films is the use of layered coatings such as a soft film of lead, ion plated onto a film of a hexagonal close packed metal such as cobalt on a hard substrate (Teer and Arnell, 1974). The lead provides the low shear strength surface and the cobalt prevents any excessive wear of the substrate when the lead film is penetrated. In a similar way gold can be ion plated on rhodium or platinum to improve wear resistance of titanium alloys (International R&D, 1981), and aluminium can be ion plated on a hard aluminium oxide multi-layered coating on a steel substrate using a pulsed ion plating technique (Ahmed and Teer, 1982; Ahmed, 1986a) (Figure 105). It has also been reported by Christy (1981) that a combination of ion plated gold and sputtered MoS_2 on harmonic drive gears usually give good lifetime and performance. Furthermore, in an attempt to improve the tribological properties of solid film lubricants, Kloos *et al.* (1981) used a thermionically assisted triode discharge to increase the ionization efficiency of the gas discharge during the ion plating of different soft metallic films. Different film combinations such as CuPb, CuIn, CuAgPb and CuAgIn have been used and the results showed a general improvement in structures so that the coatings were able to withstand high contact pressures without break-up because of the higher cohesive strength of the coatings as compared to coatings prepared in

Figure 105. Aluminium–aluminium oxide multi-layered coating using pulsed ion plating. (N. A. G. Ahmed, unpublished results, 1984)

diode ion plating. However, depending on the coating materials, the frictional properties vary. Kloos *et al.* (1981) reported that lead tended to have a bad influence on the coefficient of friction when the coating is used in a normal atmosphere since oxidation of lead can occur, resulting in a heavy wear.

7.4 OPTICAL COATINGS

Vacuum evaporation and sputtering have been used regularly to produce good optical coatings for a wide range of applications including solar energy conversion, infra-red filters and laser mirrors (Green, 1983; Holland, 1956). Such coatings should be durable, stress-free, transparent to the wavelengths in which they are required and stable at operating temperature (Macleod, 1969). Optical properties of thin films are usually influenced by many of the deposition parameters such as pressure, evaporation rate, substrate temperature, the starting material in the crucible, etc. (Ritter, 1981). Optimization of different parameters should give reproducible properties. However, one of the major problems in coatings deposited by the above mentioned techniques is the open columnar grain structure with a packing density of less than unity. Coating with such a structure will absorb water vapour upon venting to the atmosphere and therefore will have an unpredictable refractive index, which will drift with the relative humidity of the operating environment (Holland, 1956). Another problem with such coatings is the abrupt interface with little or no interdiffusion and, if such interdiffusion were to occur after deposition, then severe shifts in the optical properties should be likely (Macleod, 1979); some of these problems such as packing density can be solved by heating the substrate. Yet for some materials the packing density will be still less than unity at 300°C (Reale, 1978). There is, however, another satisfactory approach to eliminate the need for heating the substrate and that is by ion bombardment, which adds activation energy to the growing film thus disrupting the columnar growth (see Section 6.2). It has been reported by Martin *et al.* (1984) that significant improvements in the optical properties of ZrO_2 have been made following the irradiation of the growing film with low energy oxygen ions. It has also been reported by many other workers that ion plating appears to be a possible process for optical film deposition at room temperature (Bishop and Howson, 1980). This is important in the case of ion plating on flexible plastic materials such as PET (ICI Melinex), where the substrate is greatly influenced by temperature rise (Ridge, 1981, 1982). On the other hand, Henderson (1977) has reported encouraging results for ion plating zinc sulphide film on germanium substrate for infra-red anti-reflection applications. Reid *et al.* (1979) have also found that the abrasion resistance of zinc sulphide films was further improved by heating the germanium substrate to 150° and enhancing the discharge with a third electrode which permits the increase of discharge current at lower gas pressure thus yielding denser structures. Furthermore, Gautherin and Weissmantel (1978) have used successfully a system termed 'High Vacuum Ion Plating', in which an external ion source was used and ions could be extracted into the ion plating chamber during conventional evaporation to produce diamond-like structure carbon films with good optical properties. Howson *et al.* (1979, 1981) used r.f. bias ion plating with an electron beam gun to deposit TiO_2 films with a refractive index close to 2.5 on a plastic sheet of PET, and Si_3N_4 films with a refractive index of 2 at room temperature on a glass substrate. Figure 106 shows the influence of the r.f. power on the refractive index of TiO_2 at different oxygen pressures. In this case the evaporation rate has to be balanced carefully with the residual pressure of oxygen and the r.f. power applied to the substrate. Sherbiney and Aboukhashaba (1983) have successfully deposited indium and indium tin oxide on glass substrates using r.f. reactive ion plating without direct substrate heating. Indium-tin oxide coatings are usually used as infra-red reflecting surfaces for solar heat mirrors (Howson *et al.*, 1979). For continuous production of optically transparent and electrically conductive coatings on flexible plastic substrates, Ridge *et al.* (1981) designed a prototype r.f. reactive ion plating industrial system using a planar

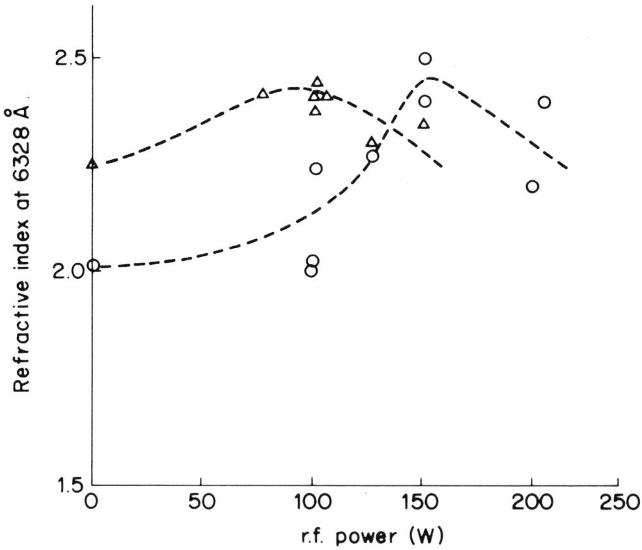

Figure 106. The influence of r.f. power on the refractive index of r.f. implanted TiO_2 films at different oxygen pressures of (○) 0.03 Pa a nd (△) 0.07 Pa (from Howson *et al.*, 1984). (*Reprinted with permission from Vacuum vol. 34, ©1984 Pergamon Journals Ltd*)

magnetron sputtering source. Films of In–10% Sn oxide on plastic with 90% transmission and 400 Ω cm^{-2} electrical conductivity have been produced by controlling precisely the sputtering rate and the mass flow rate of oxygen into the system (Ridge, 1981, 1982). The introduction of an r.f. bias to the substrate was also shown to enhance the reaction process and to relax the degree of control required to attain the highest conductivity coatings.

Ion plating has also been used to produce In_2O_3 films doped with fluorine for solar energy control (Howson *et al.*, 1981), SiO_2 films on plastic for ophthalmic lenses (Suzuki *et al.*, 1981; Suzuki and Howson, 1983) and to form lamp reflectors by depositing aluminium onto lacquered steel or plastic substrates (Plaisted, 1978). Quartz, glass or SiO_2 were also ion plated onto mirror surfaces to give abrasion resistance and solar cell encapsulation by forming pinhole-free transparent coatings (International R&D, 1981).

7.5 ELECTRICAL APPLICATION

Ion plated metal coatings have found widespread applications in electrical contacts (Sims, 1976). Heinz and Kienel (1977) have used ion plating to deposit noble metal coatings such as platinum and gold on reed blade contacts to increase their life. An automatic rapid loading and unloading reed blade holder was designed for such a purpose (Jannick *et al.*, 1979). It has also been reported that aluminium can be soldered after being ion plated with copper which also offers corrosion protection (International R&D, 1981). Ion plating metal layers such as Cu, Au and Al on ceramics has been used to manufacture microwave striplines and hybrid circuits (Delcea *et al.*, 1979). Chin and Elsver (1974) reported the use of reactive ion plating to deposit silicon nitride (Si_3N_4) coatings on n- and p-type 85 Si–15 Mo alloys. These alloys are used as thermoelectric couples and Si_3N_4 coating was applied to improve the service lives of deep-space power sources. Si_3N_4 has been reported to be a useful material in inhibiting the rate of evaporation and electrical shorting in thermoelectric devices (Galasso *et al.*, 1972). In other applications indium has been ion plated onto semiconductor devices for ohmic contacts. Pinhole-free Li-doped ZnSe thin films were produced by r.f. ion plating with a minimum electrical resistivity (Gordon and Hamah, 1974). However, on exposure to atmosphere these films become unstable. Thin film capacitors were also formed as an Al–aluminium nitride–Al structure using r.f. reactive ion plating in a nitrogen discharge. The coating showed very good adhesion and thermal stability (Mangalavaj *et al.*, 1982). In recent years, there have also been many research and development investigations on carbon fibres for military and civilian applications (Report, 1984; Murday and Dominguez, 1983). The fact that carbon fibres have very good strength to weight properties has made them attractive to the aircraft industry and at the present time fibre/epoxy composites are being used for the skin of the high performance AV-8B and F-18 aircraft. However, the poor electrical conductivity of the carbon fibre represent a real problem, particularly where electromagnetic interference (EMI) shielding by the composite aircraft skin is important. Many attempts have been made to deposit an aluminium coating on carbon fibres to reinforce the fibres and to increase the electrical conductivity (Aggour *et al.*, 1977). One problem associated with the aluminium is the poor wettability on carbon fibre and the production of unfavourable carbide during fabrication at high temperature (Baker *et al.*, 1971). In order to deposit an adherent aluminium coating on high strength carbon fibres, Ohsaki *et al.* (1977) have used the ion plating technique with some success. The author (Ahmed, 1985b) has also used ion plating to deposit thick aluminium coating on carbon fibres with electrical conductivity of 20–60 $m\Omega\ cm^{-2}$. By controlling the process parameters, such as substrate bias and current density, an adherent aluminium coating of $10 \mu m$ thick was successfully deposited on carbon fibres. Figure 107 shows the structure of the aluminium coating in a fractured sample. Figure 108 shows a comparison between uncoated and aluminium coated carbon fibre strips.

7.6 HIGH TEMPERATURE COATINGS FOR GAS TURBINE BLADES

Over the past two decades, there has been a growing interest in the coating of gas turbine blades with high temperature coatings which can withstand high stresses in a highly corrosive and hot environment of oxygen, sulphur and chlorine-containing gases without premature failure (Restall, 1979, 1981). The

Figure 107. Scanning electron micrograph of a fractured section in aluminium ion plated carbon fibres (N. A. G. Ahmed, unpublished results, 1985)

Figure 108. Comparison between uncoated and aluminium ion plated carbon fibre strips (N.A.G. Ahmed, unpublished results, 1986)

main reason for needing such coatings is that turbine rotor blades which are made of nickel or cobalt base superalloys cannot offer good protection against hot gas environments. Protective aluminide coating has been used on these superalloys to minimize oxidation and corrosion damage during service (Jansson and Rieck, 1967; Goward, 1970). However, on long runs, failure of the coating will occur due to spallation of the protective alumina film during thermal cycling and depletion of aluminium through interdiffusion between the coating and substrate to a level where a stable protective Al_2O_3 scale could not be maintained (Goward, 1970). To extend surface stability or oxidation resistance, the alloy is simply provided with a larger reservoir of aluminium. Nickel–aluminium intermetallic phase is usually formed and it has been found that the most suitable phase for protection is β(NiAl), which has a rather wide range of stoichiometry (Goward, 1970). There is another problem associated with the aluminide coating and that is in gas turbine engines which operate close to the sea (such as in Navy aircraft or maritime patrol boats), the combination of sea water and hot gases accelerate the corrosion and attack the aluminide coating and the underneath surfaces of the turbine blades (Conde, 1972; Hancock, 1968; Hart and Cutler, 1973) (Figure 109). Therefore, improved blade coatings are required to resist hot corrosion, oxidation and thermal fluctuations due to operating at different temperatures. It has been found that the most suitable coating to provide such protection is an overlay coating of M-Cr-Al-Y alloy

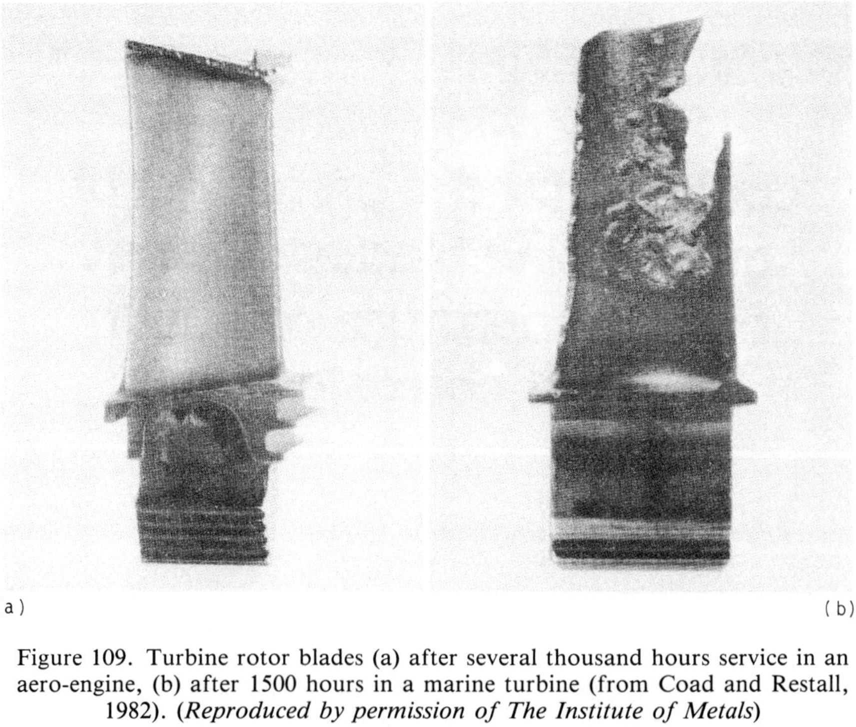

(a) (b)

Figure 109. Turbine rotor blades (a) after several thousand hours service in an aero-engine, (b) after 1500 hours in a marine turbine (from Coad and Restall, 1982). (*Reproduced by permission of The Institute of Metals*)

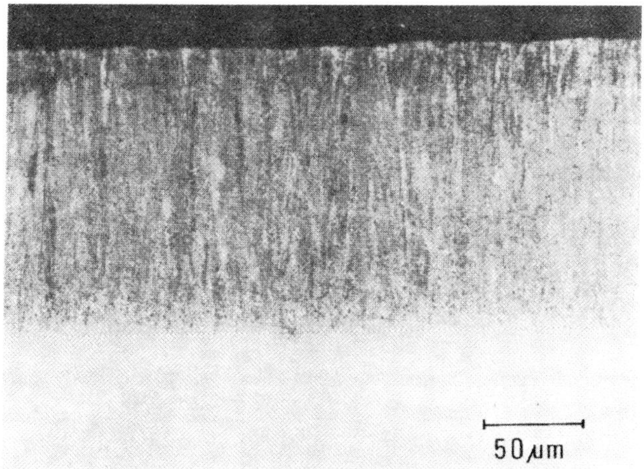

Figure 110. Optical micrographs of part of a NiCrAlTi coated Ni alloy turbine blade (from Coad and Restall, 1982). (*Reproduced by permission of The Institute of Metals*)

where M is Ni, Co, Fe or Ni + Co (Figure 110). This coating has been deposited for many years by electron beam vacuum evaporation process and plasma spraying (Talboon *et al.*, 1969). However, because of the problems of suppressing the columnar structures even at high temperature deposition, ion plating (Bunshah, 1984), sputtering (Teer, 1983) and low pressure plasma deposition (Steffans, 1980) have been used to deposit overlay coatings with good corrosion protection. There are several forms of M-Cr-Al-Y which are important for practical applications and including FeCrAlY (Chambers and Carmichael, 1971), CoCrAlY and NiCrAlTi (Restall, 1979). CoCrAlY has been used with the following nominal compositions (wt %): Co–20% Cr– 9% Al–0.3% Y, and Co–25% Cr 9% Al–0.3% Y. NiCrAlTi can also be used with the following nominal compositions: Ni–37% Cr–3% Al–2% Ti, and Ni–35% Cr–8% Al–3% Ti. Coad and Restall (1982) reported the use of sputter ion plating to deposit 80 μm thick NiCrAlTi and CoCrAlY which provided excellent corrosion resistance to the nickel alloy substrate even after 1200 hours exposure to hot corrosive gases, compared to 100 hours for aluminide and other coatings.

In general, the overlay coatings have several advantages over diffusion aluminide coatings. The latter lose their effectiveness at higher temperatures due to interactions with the substrate material (Restall, 1981). The composition and properties of the overlay coatings can also be modified for each specific application (NAVESC, 1976: Lang, 1983). However, with the potential future use of synthetic fuels, considerable research is being carried out to modify such overlay compositions for the different corrosion environments as well as against erosion from the particulate matter in those fuels (Restall, 1981).

Chapter 8
Conclusions

8.1 ADVANTAGES AND LIMITATIONS OF ION PLATING

The ion plating process has several important advantages over other coating methods; these are:

(1) Dense thick coatings can be deposited with excellent adhesion at low substrate temperature.
(2) Good coverage occurs on complicated shapes due to the excellent throwing power.
(3) By use of suitable evaporation sources, complex alloy coating can be deposited with good control of composition and high deposition rates.
(4) Excellent surface finish can be obtained.
(5) Thorough cleaning of the substrate is not always necessary as this can be carried out during the sputter cleaning cycle prior to deposition.
(6) Metals, plastics and ceramic substrates can be coated.
(7) Enhanced reactivity occurs which encourages formation of compounds having the correct chemical stoichiometry.
(8) No hydrogen embrittlement occurs when producing aluminium coatings.
(9) The process is clean and pollution free with no effluent requiring disposal.

The present limitations of the ion plating process are:

(1) The necessity of clamping the sample to a high voltage electrode.
(2) The difficulty in controlling the wide range of ion and neutral energies.
(3) The need to work in a relatively high gas pressure to achieve good throwing power.
(4) The low output of ion plating machines makes the process more expensive than others such as electro-plating.
(5) The difficulty in coating the inside of hollow tubes such as missile shells.

To facilitate the special features of ion plating, Table 8 shows a comparison between ion plating, electro-plating, vapour deposition and sputtering.

8.2 THE FUTURE OF ION PLATING

As mentioned in Chapter 7, the development of a wide range of new equipment designed for a particular application has led in itself to a further new applications. Therefore, ion plating has proved to be not only a laboratory

Table 8. Comparison of Various Deposition Methods

Characteristic	Electroplating	Vacuum Deposition	Ion Plating	Sputtering
Nature of process	Electrochemical	Vacuum evaporation	Vacuum plasma	Vacuum plasma
Materials to be coated	Metals	Metals, some compounds	Metals, alloys, some compounds	Metals, alloys, compounds, ceramics, polymers
Energy of deposited particles	Can be high	0.1–1 eV	1–1000 eV	1–10 eV
Nature of film	May contain pores, can blister	May not be uniform	Dense, pore-free, does not blister	Dense, fairly pore free
Deposition rates	Low to high	0.1–75 μm/min	0.1–25 μm/min	0.01–2 μm/min
Adhesion of film to substrate	Fair	Poor	Excellent	Good
Type of interface	Sharp	Sharp unless thermally diffused	Diffuse or graded	Relatively sharp
Throwing power	Good for complex objects, poor in blind holes	Poor	Excellent on complex objects but slightly poor in blind holes	Good on complex objects but poor in blind holes
Uniformity of film	Fairly uniform on flats, can have non-uniformity on edges	May vary	Good uniformity	Fairly good uniformity
Film purity	May contain bath salts and gaseous inclusions	Purity of starting material	Purity of starting material	Purity of starting material
Density of deposited film	Low	Slightly low	High	High
Substrate heating	No	Yes, normally	Yes or no	Not generally
Film thickness control	Fair	Poor	Excellent	Excellent
Size of object to be coated	Limited by size of bath and power supply	Limited by size of vacuum chamber	Limited by size of vacuum chamber	Limited by size of vacuum chamber
Equipment required	Chemical baths, controls and power supply	Vacuum chamber and heat source	Vacuum chamber, heat source and high voltage supply	Vacuum chamber and d.c. or r.f. power supply
Ecology problem	Chemical handling, vapours and disposal	None	None	None

process but a process that can compete industrially with other coating processes such as electro-plating. The production of large scale ion plating equipment for the aircraft, automobile and tool industry has demonstrated that ion plating will penetrate gradually into the field of electroplating in the near future, particularly when cadmium will be banned for its toxicity. However, the research and development on a laboratory scale is still a critical step towards a further successful industrial application, in particular for coatings that are difficult to achieve by other coating methods. Furthermore, the versatility of ion plating technology will assure new applications and markets in the near future. As Professor Bunshah stated in 1974 (Bunshah, 1974) 'looking into the crystal ball, it is not difficult to see that materials requirements for the energy related applications, for higher productivity in machining and forming, for longer life of components, are spurring the development of coating technologies at a rapid tempo'.

Indeed, ion plating technology has been the front runner among the many other coating technologies for the last decade and it is expected to be so in the next decade.

The following is a list of some potential further applications of ion plating where research and/or development has shown promising results on a laboratory scale and in limited industrial applications.

1. Corrosion Protection

Improvements of the characteristics of ion plated aluminium will give better corrosion protection (Ahmed, 1984). Aluminium–zinc has shown promise in the sacrificial protection of steel in a salt spray environment, but is less good in other applications (Abu Zeid, 1982).

Other alloy coatings such as aluminium–tin, aluminium–copper, aluminium–magnesium (Nowak *et al.*, 1985), and aluminium–aluminium multi-layered coatings (Ahmed, 1986a) are also under investigation. For high temperature corrosion protection, coatings of alumina, zirconia, silica and new MCrA1Y alloy are under consideration (Cook, 1983).

2. Abrasion, Wear and Erosion Resistance

New coatings of carbides, nitrides, oxides and other multi-layered coatings are expected to be used on common steel and high speed steel to increase further the tool performance. Erosion resistant coatings such as chromium, molybdenum, tungsten, tantalum and titanium are expected to be used on a wider scale on gun steel to protect the bores of larger calibre tank guns (Jones *et al.*, 1975; Gipson, 1981).

3. Electrical Coatings

These include: aluminium, gold and nickel for lead frames (Plaisted, 1978; Smith, 1974); aluminium electrodes for glass vacuum devices (Young and Hurley,

1979); copper, gold, and aluminium for microwave applications (Delcea, 1979); aluminium or copper on carbon fibres and plastic to reduce EMI shielding (Ahmed, 1985b).

4. Optical Coatings

Wider use of indium-tin oxide (as optically transparent and electrically conductive film) may be expected; also titanium oxide, zinc oxide, copper, silver and aluminium on plastic for energy conversion (Bishop and Howson, 1980; Reid *et al.*, 1979; Howson *et al.*, 1979; Suzuki *et al.*, 1981; Gautherin and Weissmantel, 1978). It is also expected that other ion plated coatings, such as zinc selenide (ZnSe), tantalum oxide (Ta_2O_5) and yttrium oxide (Y_2O_3) will be used as anti-reflection, infra-red optical filters and for integrated optical applications (Varasi *et al.*, 1986).

5. Plastic

Ion plating is expected to replace vacuum coating for metallizing plastic. Results from an experimental ion plating web coater indicates the possibility of depositing indium-tin oxide coating and other coatings on moving plastic web (Ridge *et al.*, 1981). Ion plating is also expected to be used in conjunction with electro-plating to deposit conductive films such as chromium or copper on plastic prior to depositing thicker films by electro-plating. Possible markets are the automobile, solar, packaging and electronics industries (Dini, 1985).

6. Ceramics

Ion plating metals on ceramics will enable easier joining by soldering or brazing (International R&D, 1981).

7. Foils

The preparation of self-supported ion plated foils such as carbon, metals and alloys with mechanical properties equal or superior to that conventionally produced by wrought material (Tolfree *et al.*, 1979) will be possible.

8. Refurbishment

Refurbishing worn components such as pistons in aircraft engines could be done by ion plating sufficiently thick coatings (White, 1973).

9. X-ray generation

Aluminium and magnesium can be plated on copper as targets to produce soft x-rays (Teer, 1975).

10. Tin plate substitute

Aluminium or other coatings may be plated onto steel strip to replace tin plate in the steel industry (Schiller *et al.*, 1974).

11. Decoration

Ion plating of a yellow-gold coloured titanium nitride coating is expected to replace gold in jewellery, watch cases, straps, cutlery and other decorative applications (Moll and Daxinger, 1977).

12. Lubrication

Development of molybdenum sulphide and other coatings for dry lubrications (Dimigen and Hubsch, 1983) is under way.

8.3 ECONOMICS OF THE ION PLATING PROCESS

The economics of the ion plating process depends mainly on the capital investment in the equipment and other expenses such as additional units for cleaning and preparing the items for coating. This, in turn, controls the price of the coating per item and thus makes the process cost competitive with other processes, such as electroplating. In general, there are two concepts for the economics of the process. The first is the go/no go concept where new types of coatings with improved or superior properties can only be supplied by the ion plating process or cannot be economically supplied by other coating processes (Bunshah, 1984). In such a case, the customer has to bear the capital cost of the equipment and any additional plants since the superiority of the coating justifies its cost. For example, refurbishing worn components such as aircraft engine pistons may cost around 5–10 times less than installing new pistons, in addition to the fact that the coating has certain desired properties that are unobtainable otherwise (Bunshah, 1984). Another example is in coating aircraft fasteners and other parts for corrosion protection. In this case ion plating with aluminium is a desirable technique to eliminate the problem of hydrogen embrittlement which usually causes premature failure of the parts. Commercial barrel coaters (Ahmed, 1986a) can ion plate around 2000 fasteners per hour with aluminium. After taking into consideration the capital cost of the machine and other expenses, the cost of coating per fastener will be around 5–10 pence (Spencer Franklin, 1985). Although this cost is around 10 times higher than electroplating, for an expensive aircraft fastener with good corrosion protection this cost is negligible. Further example is in coating a turbine blade with MCrAlY where the cost of coating is 2–5 times less than the cost of the actual blade. In this case the cost for coating is immaterial since the desired properties are unobtainable otherwise (Bunshah, 1982).

The second concept is to make the process cost competitive with other coating techniques. For example, the automobile industry does not have strict specifications like the aircraft industry for the number of hours of corrosion protection or hydrogen embrittlement, plus the fact that car fasteners are cheap, traditionally 'disposable' and less subject to performance monitoring. In this case the only drive is the cost of coating per fastener and the throughput of the coating process. Therefore, aluminium ion plating with barrel coaters is still expensive compared to electro-plating and in order to compete it has to be designed to operate on a continuous basis. Similarly, a cutting tool may be coated with titanium nitride by PVD or CVD processes. The cost/delivery/convenience concept is then predominant (Matthews, 1985). Therefore, ion plating a £100 hob with titanium nitride at a cost of £30, and increasing its lifetime 4–5 times longer may be more economical compared to coating a £1 twist drill with titanium nitride at a cost of 30–50 pence. Although the increase in the lifetime for a low price twist drill is relatively unimportant, the titanium nitride coating allows a significant increase in speed and feed rate which may increase the productivity on the shop floor by a factor of 40% and therefore justify the price for coating (Matthews, 1985). Time used in preparing the samples such as cleaning, masking, racking and peening also plays an important factor in increasing the cost of coating. For example, time spent in cleaning soiled and rusty parts is usually longer than cleaning freshly produced parts. Other factors involved in pricing is the complexity of the coating such as the case in multi-layered coating, as well as the terms and conditions of equipment sale. Some manufacturers may offer an all inclusive turn-key operation, while others may charge extra for certain necessary plant or technology transfer. Some companies may even request royalty payments, or even an equity stake from customer companies. The customer should therefore be aware of such factors before considering the purchase of the required equipment.

REFERENCES

Abu Zeid, O. (1982). *PhD Thesis*, University of Salford, UK.

Aggour, L., Fitzer, E., Heym, M., and Ignatowitz, E. (1977). *Thin Solid Films* **40**, 97.

Ahmed, N. A. G. (1974) 'Preliminary studies of ion Plating Processes', *MSc Thesis*, University of Salford, UK.

Ahmed, N. A. G. (1974). Unpublished work.

Ahmed, N. A. G. (1980). *J. Phys. E: Scientific Instruments* **13**, 1305.

Ahmed, N. A. G. (1984). *Vacuum* **34**, 711.

Ahmed, N. A. G. (1985a). *Surfacing Journal* **16**(3), 74.

Ahmed, N. A. G. (1985b). Unpublished work on Carbon Fibres.

Ahmed, N. A. G. (1985c). UK Patent Application No. 8504458.

Ahmed, N. A. G. (1986a). *Surface Engineering* **2**, No. 2, 95.

Ahmed, N. A. G. (1985b). To be published in *Thin Solid Films*.

Ahmed, N. A. G., and Teer, D. G. (1981). *Thin Solid Films* **80**, 49.

Ahmed, N. A. G., and Teer, D. G. (1982). *Proc. 4th Int. Colloquium on Plasmas and Sputtering, Nice, France* (Paris; Societe Francais du Vide), pp. 191.

Ahmed, N. A. G. and Teer, D. G. (1984). *J. Phys. E: Scientific Instruments* **17**, 411.

Ahmed, N. A. G., Colligon, J. S., and Hill, A. E. (1985). *Thin Solid Films* **129**, 255.

Ahmed, N. A. G., and Colligon, J. S. (1986). To be published in *Vacuum*.

Aisenberg, S. (1972). *Proc. Conf. Sputtering and Ion Plating, NASA Report SP5111*, 171. (NASA Lewis Research Centre.)

Aisenberg, S., and Chabot, R. W. (1973). *J. Vac. Sci. Technol.* **10**, 104.

Andersen, H. H. (1979). *Appl. Phys.* **18**, 131.
Archard, J. F., and Hirst, W. (1956). 'The wear of metals under unlubricated conditions', *Proc. Roy. Soc.* **A236**, 200.
Armour, D. G., 1986, Private Communication
Armour, D. G., Carter, G., Webb, R. P., Ingram, D. C., and Newcombe, R. (1980). *Rad. Effects Lett.* **50**, 45.
Armour, D. G., Valizadeh, H., Soliman, F. A., and Carter, G. (1984). *Vacuum* **34**, 295.
Arnell, R. D., and Soliman, F. A. (1978). *Thin Solid Films* **53**, 333.
Auwarter, M. (1960). US Patent No. 2.920.002.
Babaev, V. O., Bykov, Ju. V., and Guseva, M. B. (1976). *Thin Solid Films* **38**, 1.
Baker, A., Martin, A., and Pache, R. (1971). *Composites* **2**, 154.
Benjamin, P., and Weaver, C. (1960). *Proc. Roy. Soc.* **A254**, 163.
Berghaus, B. (1938). UK Patent No. 510, 993.
Bishop, C. A., and Howson, R. P. (1980). *Thin Solid Films* **72**, 341.
Bisson, E. E. (1964). *Advanced Bearing Technology*, NASA, SP–38, 259.
Bland, R. D. (1968). *Electrochemical Technology* **6**, 7272.
Bland, R. D., Kominiak, G. J., and Mattox, D. M. (1974). *J. Vac. Sci. Technol.* **11**, 671.
Blocher, J. (1982). *Deposition Technologies for Films and Coatings* Ed. R. F. Bunshah, p. 335, Noyes Publications, USA.
Boone, D. H., Strangmon, T. E., and Wilson, L. W. (1974). *J. Vac. Sci. Technol.* **11**, 641.
Borax Consolidated (Borides) Ltd. (1985). Private Communication.
Boving, H., Hintermann, H. E., and Stehle, M. (1981). *Lubr. Eng.* **37**, 534.
Boving, H., Hintermann, H. E., Begelinger, A. and de Gee, A. W. I. (1983). *Wear* **88**, 13.
Bowden, F. B., and Tabor, D. (1964a). *The Friction and Lubrication of Solids* Part I, Oxford University Press, London.
Bowden, F. B., and Tabor, D. (1964b). *The Friction and Lubrication of Solids*, Part II, Clarendon Press, London.
Brande, G. (1977). *Plating and Metal Finishing* **64**, 8.
Buhl, R., Pulker, H. K., and Moll, E. (1981). *Thin Solid Films* **80**, 265.
Bukley, D. H., and Spalvins, T. (1977). 'Effect of ion plated films on germanium and silicon on friction, wear and oxidation of 521000 bearing steel', *NASA Technical Note*, TM D–8436.
Bunshah, R. F. (1974). 'Physical vapour deposition of metals, alloys and ceramics', in *New Trends in Materials Processing*, American Society for Metals.
Bunshah, R. F. (1981). *Thin Solid Films* **80**, 255.
Bunshah, R. F. (Ed.) (1982). *Deposition Technologies for Films and Coatings*', Noyes Publications, USA.
Bunshah, R. F. (1984). *J. Vac. Sci. Technol.* **B2(4)**, 789.
Bunshah, R. F., and Deshpandey, C. V. (1985). *J. Vac. Sci. Technol.* **A3(3)**, 553.
Bunshah, R. F., and Juntz, R. S. (1972). *J. Vac. Sci. Technol.* **9**, 1404.
Bunshah, R. F., and Schramm, R. J. (1977). *Thin Solid Films* **40**, 211.
Burns and Bradly (1967). *Protective Coatings for Metals*, 3rd edn, Reinhold, New York.
Burton, R. A. (1965). *Lubr. Eng.* **21**, 227.
Carter, G., and Armour, D. G. (1981). *Thin Solid Films* **80**, 13.
Carter, G., and Grant, W. A. (1976). *Ion Implantation of Semiconductors*, Edward Arnold, London.
Chabicovsky, R, Pulker, H. K., and Schadler, W. (1970). *Vak. Techn.* **19**, 101.
Chambers, D. L., and Carmichael, D. C. (1971). *Research/Development* **22**, 32.
Chapman, B. (1980). *Glow Discharge Processes, Sputtering and Plasma Etching*, Wiley International Ltd, New York.
Chaudri, R. M., and Oliphant, M. L. (1932). *Proc. Roy. Soc.* **A137**, 662.
Cheavallier, J., Chabert, J. P., and Spitz, J. (1981). *This Solid Films* **80**, 263.
Chin, J., and Elsver, N. B. (1974). *J. Vac. Sci. Technol.* **12**, 821.
Chopra, K. L. (1969). *Thin Film Phenomena*, McGraw-Hill, New York.
Christy, R. I. (1981). *Thin Solid Films* **80**, 289.
Clarke, P. J. (1977). *J. Vac. Sci. Technol.* **14**, 141.
Coad, J. P., and Restall, J. E. (1982). *Proc. Conf. on Ion Assisted Surface Treatments, Techniques and Processes, 14 September 1982*, The Metal Society, UK, p. 13.1.
Conde, J. F. G. (1972). *Proc. Specialist Conf. on High Temperature Corrosion of Aerospace Alloys, Lyngby, Denmark*, AGARD-CP-120, pp. 203.
Cook, K. E. (1983). 'Protective coatings for engineering components', ERA Report No. 83–0135, ERA Technology, UK.

Cuomo, J. J. (1982). *Proc. 25th Ann. Tech. Conf. Soc. Vac. Coaters*, p. 48.

Cuomo, J. J., Harper, J. M., Guarnieri, C. R., Yee, D. S., Attanasio, L. J., Angilello, J., Wu, C. T., and Hammond, R. H. (1982). *J. Vac. Sci. Technol.* **20**, 349.

Davis, W. D., and Vanderslice, T. A. (1963). *Phys. Rev.* **131**, 219.

Delcea, B. L. (1978). *PhD Thesis*, University of Salford, UK.

Delcea, B. L., Cavalera, A., Raicu, B., and Dina, G. (1979). *Proc. Conf. Ion Plating and Allied Techniques, IPAT 79*, CEP Consultants, Edinburgh, p. 161.

Delcea, B. L., and Dunare, B. A. (1983). *Proc. Int. Ion Engineering Congress, ISIAT 83 and IPAT 83*, Japan, p. 1313.

Deshpandey, C., and Holland, L. (1982). *Thin Solid Films* **96**, 265.

Dimigen, H., and Hubsch, H. (1983). *Philips Tech. Rev.* **41**, 186.

Dini, J. W. (1985). *Plating and Surface Finishing*, p. 48.

Dirks, A. G. and Leamy, H. J. (1977). *Thin Solid Films* **47**, 219.

Dobrev, D., and Marinov, M. (1973). *C. R. Acad. Bulg. Sci.* **26**, 231.

Dorodnov, A. M., and Petrosov, B. A. (1981). *Sov. Phys. Tech. Phys.* **26**, 304.

Dorodnov, A. M., and Miroshkin, S. I. (1981). *High Temp.* **18**, 821.

Drugresteyn, M. J., and Penning, F. M. (1940). *Rev. Mod. Phys.* **12**, 87.

Dugdale, R. A. (1966). *J. Mater. Sci.* **1**, 160.

Dugdale, R. A. (1977). *Proc. Conf. Ion Plating and Allied Techniques, IPAT 77*, CEP Consultants, Edinburgh, p. 179.

Eltoukhy, A. H., and Green, J. E. (1980). *J. Appl. Phys.* **51**, 4444.

Enomoto, Y., Tsuya, Y., Yaguchi, G., Ohkami, T., and Takagi, R. (1979). *Proc. Conf. Ion Plating and Allied Techniques, IPAT 77*, CEP Consultants, Edinburgh, p. 29.

Etzkorn, H. W., Littmark, U., and Kirschner, J. (1980), in *Symposium on Sputtering*, Ed. V. Verga, G. Betz and F. P. Viehback, Vienna, Inst. fur Allgemeine Physik, p. 542.

Fager, D. N., and Spuir, W. F. (1970). 'Solid cadmium embrittlement: titanium alloys', *Corrosion* **26**, 10.

Fager, D. N., and Spuir, W. F. (1971). *Corrosion–NACE* **27**, 72.

Fan, J. C. C., Bachner, F. J., Foley, G. H., and Zarracky, P. M. (1974). *Appl. Phys. Lett.* **25**, 693.

Fancey, K. S., and Beynon, J. (1984). *Vacuum* **34**, 591.

Fannin, E. R. (1977). McAir Report No. 77–012.

Fannin, E. R., and Steube, K. E. (1974). McAir Report 74–008.

Finklin, E. (1971). *Wear* **18**, 231.

Fishter, R. (1978). Pratt and Whitney Aircraft Group Report No. FR9865.

Fleischer, W., Schulze, D., Wilbery, R., Lunk, A., and Schiode, F. (1979). *Thin Solid Films* **63**, 347.

Forant, P. R., and Habdanian, M. (1978). *Proc. 21st Ann. Conf. Soc. Vacuum Coaters*, p. 40.

Francis, G. (1956). *Handbuch der Physik*, Ed. S. Flugge, Vol. 22, p. 53, Springer-Verlag, Berlin.

Franks, J. H., Stuart, P. R., and Withers, R. B. (1979). *Thin Solid Films* **60**, 231.

Galasso, F., Kuntz, U., and Craft, W. J. (1972). *J, Am. Ceram. Soc.* **55**, 431.

Gautherin, G., and Weissmantel, Chr. (1978). *Thin Solid Films* **50**, 135.

Gilbert, L. R., Messier, R., and Roy, R. (1978). *Thin Solid Films* **54**, 149.

Gipson, I. G. (1981). *Thin Solid Films* **53**, 27.

Glang, R. H., Holmwood, R. A., Kurtz, J. (1970). In *Handbook of Thin Film Technology*, Ed. L. Maissel and R. Glang, McGraw-Hill, New York, p. 2–1.

Glasser, P., and Szigeti, E. (1983). *Vacuum* **33**, 31.

Glue, D. R., Sherward, J. A., Young, W. J., and Gipson, I. P. (1980). *Proc. Inst. Metal Conf. Environmental Degradation of High Temperature Materials*, Paper 6/38, Douglas, IOM.

Goode, A. R., and Burden, M. St. J. (1979). *Vacuum* **29**, 9.

Gordashnik, K. Z., Gorokhovsky, V. I., and Uryukov, B. A. (1983). *Proc. Int. Ion Engineering Congress, ISIAT 83 and IPAT 83*, Kyoto, Japan, p. 956.

Gordon, J., and Hamah, J. J. (1974). *J. Vac. Sci. Technol.* **11**, 43.

Gouault, J., Faure, A., Hubin, M., Jontel, J., and Mesnier, D. (1971). 'Mechanical behaviour and electrical isolation of alumina films evaporated onto steel substrates by electron beam evaporation', *Le Vide*, No. 154, p. 178.

Goward, G. W. (1970). *J. Metals* **22**, 31.

Green, J. E. (1983). *CRC Critical Review of Solid State and Material Science* **11**, 47.

Green, J. E., and Zilko, J. L. (1978). *Surface Sci.* **72**, 109.

Green, M. (1982). *Solar Cells; Operating Principles, Technology and System Applications*, Prentice-Hall Inc., Englewood Cliffs, NJ.

Halling, J. (1979). *Principles of Tribology*, MacMillan Press Ltd, London.

Hammer, T. (1970). *Research/Development* **21**, 53.
Hancock, P. (1968). *Corrosion of Alloys at High Temperatures in Atmospheres Consisting of Fuel Combustion Products and Associated Impurities—A Critical Review*, HMSO, London.
Hanni, W., and Hintermann, H. E. (1979). *Proc. Int. Chronometry Congress (Swiss Society of Chronometry), Buren, Switzerland*, Vol. 10, p. 455.
Harper, J. M., Cuomo, J. J., Gambino, R. J., and Kaufmann, H. R. (1984) in *Ion Bombardment Modification of Surfaces; Fundamentals and Applications*, Ed. O. Auciello and R. Kelly, Elsevier Science Publishers, Amsterdam, p. 127.
Harris, N. S. (1983). *Vacuum Technology*, Edwards School of Vacuum.
Harrower, G. A. (1955). *Rev. Sci. Instrum.* **26**, 850.
Hart, A. B., and Cutler, A. J. B. (Ed.) (1973). *Conf. Deposition and Corrosion in Gas Turbines, CEGB*, Applied Science, London.
Hatschek, R. L. (1983). *American Machinist*, Special Report, 752, p. 129.
Hatto, P., and Teer, D. G. (1986). *Vacuum* **36**, 67.
Heavens, O. (1950). *J. Phys. Rad.* **11**, 355.
Hecht, R. J., and Mullaly, J. R. (1975). *J. Vac. Sci. Technol.* **12**, 836.
Heinz, B. and Kienel, G. (1977). *Proc. Conf. Ion Plating and Allied Techniques, IPAT 77*, CEP Consultants, Edinburgh, p. 13.
Helevirta, P., and Kettumen, P. (1986). *Proc. Int. Conf. Metallurgical Coatings, ICMC 84, San Diego, USA*. To be published in *Thin Solid Films*.
Heller, J. (1970). *Thin Solid Films* **17**, 163.
Helmersson, U., Johanssen, B. O., Sundgren, J. E., Hentzell, H. T. G., and Billgrem, P. (1985). *J. Vac. Sci. Technol.* **A3(2)**, 308.
Helmersson, U., Johanssen, B. O., Sundgren, J. E., Hentzell, H. T. G., and Billgrem, P. (1986). To be published in *J. Vac. Sci. Technol.*
Henderson, E. (1977). *IPAT 77*, CEP Consultants, Edinburgh, p. 122.
Hintermann, H. E. (1980). *Tribology Int.* **13**, 267.
Hintermann, H. E. (1981). *Thin Solid Films* **84**, 215.
Hintermann, H. E. (1984). *J. Vac. Sci. Technol.* **B2(4)**, 816.
Hintermann, H. E., Boving, H., and Hanni, W. (1978). *Wear* **48**, 225.
Hirsch, E. H., and Varga, I. K. (1980). *Thin Solid Films* **69**, 99.
Hmiel, A. F. (1985). *J. Vac. Sci. Technol.* **A3(3)**, 592.
Hoffman, D. W., and Gaerttner, M. R. (1980). *J. Vac. Sci. Technol.* **17**, 425.
Holland, L. (1956). *Vacuum Deposition of Thin Films*, Chapman and Hall Ltd, London.
Holland, L. (1965). *Thin Film Microelectronics*, John Wiley & Sons Inc., London.
Holland, L. (1974a) in *Science and Technology of Surface Coating*, Ed. B. N. Chapman and J. C. Anderson, Academic Press, London.
Holland, L. (1974b). *Vacuum Manual*, Ed. L. Holland, W. Steckelmacher and J. Yarwood, E & F N Spon, London.
Holland, L. (1975). *Thin Solid Films* **27**, 185.
Holland, L. (1977). *J. Vac. Sci. Technol.* **14**, 5.
Holland, L. (1978). *Vacuum* **28**, 437.
Houston, J. R., and Uhl, J. E. (1971). Sandia Corporation Report SC-RR-71,122.
Howatson, A. M. (1975) *An Introduction to Gas Discharge*, Pergamon Press, Oxford.
Howson, R. P., Avarissiotis, J. N., Ridge, M. I., and Bishop, C. A. (1979). *Proc. Conf. Ion Plating and Allied Techniques, IPAT 79*, CEP Consultants, Edinburgh, p. 84.
Howson, R. P., Ridge, M. I., and Bishop, C. A. (1981). *Thin Solid Films* **80**, 137.
Howson, R. P., Suzuki, K., Bishop, C. A., and Ridge, M. I. (1984). *Vacuum* **34**, 291.
Hu, S. M., Gregor, L. V., and Maissel, L. I. (1967). *IBM Tech. Dis. Bull.* **10**, 100.
Hurley, R. E. (1981). *Thin Solid Films* **86**, 241.
Hutchinson, D. A. (1962). *Adv. in Mass Spectrometry, Proc. Conf, Oxford* **2**, 527.
Industrial Finishing (1978). 'Ion vapour deposition (IVD) aluminium increases aircraft life', p. 28.
International R&D Co. (1981). 'Surface coatings by ionic bombardment', *The Production Engineer*, p. 37.
Ion Plating and Allied Techniques, IPAT 77 (1977) CEP Consultants, Edinburgh.
Ion Plating and Allied Techniques, IPAT 79 (1979) CEP Consultants, Edinburgh.
Ion Plating and Allied Techniques, IPAT 81 (1981) published in *Thin Solid Films* **80**, Nos. 1, 2, 3.
Ion Plating and Allied Techniques, IPAT 83 (1983) Kyoto, Japan.
Ion Plating and Allied Techniques, IPAT 85 (1986) Germany, published in *Vacuum* **36**, Nos. 1, 2, 3.

Jacobson, E. E., Nimmagadda, R., and Bunshah, R. F. (1979). *Thin Solid Films* **63**, 333.
Jacobson, S., Johnsson, B., and Sundquist, B. (1983). *Thin Solid Films* **107**, 89.
Jannick, R. F., Heiden, C. R., and Guttensohn, A. E. (1979). *J. Vac. Sci. Technol.* **11**, 535.
Jansson, M. M., and Rieck, G. D. (1967). *Trans. AIME* **239**, 1372.
Jones, R. H., Moss, R. W., McClanahan, E. D., and Butts, H. L. (1975). 'The sputter deposition and evaluation of tungsten and chromium coatings for use in weapon components', Technical Report No. AD A024109, US Army.
Kaufman, H. R. (1963). *NASA Tech. Note* TND-585.
Kaufman, H. R. (1974). In *Advances in Electronics and Electron Physics*, Ed. L. Marton, Academic Press, New York, vol. **36**, p. 265.
Kaufman, H. R. (1978). *J. Vac. Sci. Technol.* **15**, 272.
Kelly, R. (1980). *Symposium on Sputtering*, Ed. P. Verga, G. Betz and F. P. Viehback, Vienna, Inst. fur Allgemeine Physik, p. 390.
Kennedy, K. D., Schnermann, G. R., and Smith, H. R. (1971). *Research/Development* **22**, No. 11, p. 40.
Kloos, K. H., Broszeit, E., and Gabriel, H. M. (1981). *Thin Solid Films* **80**, 307.
Kominiak, G. K., and Mattox, D. M. (1977). *Thin Solid Films* **40**, 141.
Komiya, S., Umezu, N., and Hayashi, C. (1974). *Proc. 6th Int. Conf. Special Melting and Metallurgical Coatings, San Diego, USA*, p. 205.
Komiya, S., Umezu, N., and Hayashi, C. (1979). *Thin Solid Films* **63**, 341.
Konig, H. C. W. (1982). *Industrie Anzeiger* **27**, No. 53, p. 10.
Krageliski, I. V. (1965). *Friction and Wear*, Butterworths, London.
Lamont, L. T., and Lang, A. (1970). *J. Vac. Sci. Technol.* **7**, 198.
Lang, E. (Ed.) (1983). *Coatings for High Temperature Applications*, Allied Science Publishers, London.
Larden, M., Buhl, R., Singer, H., Pulker, H. K., and Moll, E. (1978). *Thin Solid Films* **54**, 317.
Lau, S. S., Mills, R. H., and Muth, D. G. (1972). *J. Vac. Sci. Technol.* **9**, 1196.
Lewis, B., and Campbell, D. S. (1967). *J. Vac. Sci. Technol.* **4**, 209.
Littmark, U., and Hofer, W. O. (1980). *Nucl. Instrum. Methods* **170**, 177.
Lu C. (1978) US Patent No. 3.732.778.
Luzin, A. N., and Tyushev, A. N. (1979). *Sov. Phys.-Tech. Phys.* **24**, 1508.
Machet, J., Gadet, G., Saulnier, P., and Guille, J. (1977). *Proc. 7th Int. Vac. Congr. and 3rd Int. Conf. Solid Surfaces, Vienna*, p. 86.
Machet, J., Saulnier, P., Ezquerra, and Guille, J. (1983). *Vacuum* **33**, 279.
McDonnell Douglas Aircraft Corporation (1985). Private Communication.
Macleod, H. A. (1969). *Thin Film Optical Filters*, Adam Hilger, Bristol.
Macleod, H. A. (1979). *Proc. Conf. Ion Plating and Allied Techniques, IPAT 79*, CEP Consultants, Edinburgh, p. 74.
McLoughlin, U. C. R. (1978). *AGARD, Conf. Proc.* No. 256, (*Advanced Fabrication Processes*), p. 6-1.
Maissel, L., and Glang, R. (Ed.) (1970). *Handbook of Thin Film Technology*, McGraw Hill, New York.
Maniv, S., Miner, C., and Westwood, W. D. (1983). *J. Vac. Sci. Technol.* **A1**, 1370.
Maniv, S., and Westwood, W. D. (1980). *J. Vac. Sci. Technol.* **3**, 743.
Mangalavaj, D., Radhakrishnan, M., and Balasubramania, C. (1982). *J. Phys. D: Appl. Phys.* **15**, 467.
Martin, P. J., Netterfield, R. P., and Sainty, W. G. (1984a). *J. Appl. Phys.* **55**, 235.
Martin, P. J., Sainty, W. G., and Netterfield, R. P. (1984b). *Appl. Optics* **23**, 2668.
Martin, P. J. (1986). To be published in *Vacuum*.
Matsumura, Y., Yoshioka, T., and Huang, Y. C. (1983). *Proc. Int. Ion Engineering Congress, ISIAT 83 and IPAT 83, Kyoto, Japan*, p. 1283.
Matthews, A. (1980). *PhD Thesis*, University of Salford, UK.
Matthews, A. (1984). *Proc. 1st Conf. Materials Engineering*, Ed. A. N. Bramley, Institute of Metallurgists, London.
Matthews, A. (1985). *Surface Engineering* **VI**, 93.
Matthews, A., and Teer, D. G. (1979). *Proc. Conf. Ion Plating and Allied Techniques, IPAT 79*, CEP Consultants, Edinburgh, p. 11.
Matthews, A., and Teer, D. G. (1980). *Thin Solid Films* **72**, 541.
Matthews, A., and Teer, D. G. (1981). *Thin Solid Films* **80**, 41.
Matthews, A., and Teer, D. G. (1984). *Thin Solid Films* **117**, 261.
Matthews, A., and Sundquist, H. A. (1983). *Proc. Int. Ion Engineering Congress, ISIAT 83 and IPAT 83, Kyoto, Japan*, p. 1201.

Mattox, D. M. (1963). *J. Appl. Phys.* **34**, 2493.
Mattox, D. M. (1964). *Electrochemical Techn.* **2**, 9.
Mattox, D. M. (1965). *Sandia Corp. Monograph* SC-R-65-852.
Mattox, D. M. (1966). 'Design consideration for ion plating', *Sandia Corp. Report* No. SC-R-65-997.
Mattox, D. M. (1969). 'Interface formation, adhesion and ion plating', *Trans. SAE*, p. 2175 (Review).
Mattox, D. M. (1973). *J. Vac. Sci. Technol.* **10**, 47.
Mattox, D. M. (1974a). 'Sputter deposition and ion plating technology', *AVS-Thin Film Division Monograph.*
Mattox, D. M. (1974b). *Japan. J. Appl. Phys.*, Suppl 2, Pt. 1, p. 443.
Mattox, D. M. (1978). 'Thin film adhesion and adhesive failure–A perspective' in *Adhesion of Thin Films, Thick Films and Bulk Coatings*, ASTM STP 640, Ed. K. L. Mittal, American Society for Testing and Materials, p. 54.
Mattox, D. M. (1979). *Proc. Conf. Ion Plating and Allied Techniques, IPAT 79*, CEP Consultants, Edinburgh, p 1.
Mattox, D. M. (1980). 'Fundamental processes in ion plating', presented at the 109th AIME Ann Meeting, Las Vegas, NY.
Mattox, D. M. (1982a). *Proc. 4th Int. Colloquium on Plasmas and Sputtering, Nice, France* (Paris: Societe Francais du Vide), p. 187.
Mattox, D. M. (1982b) in *Deposition Technologies for Films and Coatings*, Ed. R. F. Bunshah, Ch. 6, Noyes Publications, USA.
Mattox, D. M., and Kominiak, G. J. (1971). *J. Vac. Sci. Technol.* **8**, 194.
Mattox, D. M., and Kominiak, G. J. (1972). *J. Vac. Sci. Technol.* **9**, 528.
MIL-C-83488 (USAF) (1976). Military Specification 'Coating aluminium, ion vapour deposited'.
Miles, J. L. (1968). *J. Electrochem. Soc.* **115**, No. 12, 1255.
Military (US) Standard 1312. 'Fasteners, Test Methods'.
Minnis, E., and Sundquist, H. (1981). *Thin Solid Films* **80**, 55.
Mitchell, I. V., Williams, J. S., Smith, P., and Elliman, R. G. (1984). *Appl. Phys. Lett.* **44**(2), 192.
Miyakawa, Y. (1978). *J. JSLE (Jun Katsu)* **23**, 136.
Moll, E., and Daxinger, H. (1977). US Patent 4197175.
Morris, A. W. (1976). *Proc. AES Solar Symp, Atlanta, November 9–10.* published in *Plating and Surface Engineering*, p. 42.
Movchan, B. A., and Demchishin, A. V. (1969). *Phys. Met. Metall.* **28**, 83.
Muehlberger, D. E. (1978). *Plating and Surface Finishing* **65**, 20.
Muehlberger, D. E., and Reilly, J. J. (1985). *Proc. 19th Ann. Airline Plating and Metal Finishing Forum*, published by the Society of Automotive Engineers, p. 19.
Muehlberger, D. E., and Reilly, J. J. (1983). *Proc. 19th Ann. Airline Plating and Metal Finishing Forum*, published by the Society of Automotive Engineers, p. 127.
Munz, W. D., Hoffman, D. H., and Schuller, K. H. (1983). *Proc. Int. Ion Engineering Congress, ISIAT 83 and IPAT 83, Kyoto, Japan*, p. 1289.
Murayama, Y. (1975). *J. Vac. Sci. Technol.* **12**, 818.
Murday, J. A., and Dominguez, D. D. (1983). 'Graphitized carbon fibres: electrical conductor with high strength to volume properties', *NR Reviews*, p. 3.
Nakahara, S., and Hebard, A. F. (1983). *Thin Solid Films* **102**, 345.
Nakai, T. Ohmae, N., and Tsukizo, T. (1974). *Bull. Japan. Soc. Proc. Eng.* **8**, 17.
NASA SP5111P (1972). *Proc. Conf. Sputtering and Ion Plating, Lewis Research Centre*, Ed. B. T. Lundin.
NAVESC (1976). *Proc. 3rd US/UK Conf. on Gas Turbine Materials in The Marine Environment, Bath, UK.*
Navinsek, B., and Zabkar, A. (1986). *Vacuum* **36**, 111.
Nelson, R. S. (1977). *Proc. Int. Conf. on Ion Plating and Allied Techniques, IPAT 77*, CEP Consultants, Edinburgh, p. 32.
Nicolet, M. A. (1978). *Thin Solid Films* **52**, 415.
Nordiko Ltd, UK. (1986) *Customers' Manual for RF Sputtering Equipment.*
Nowak, W. B., Levy, M., and Chang, F. (1985). *Conf. on Applications of Ion Plating and Implantation to Materials, Atlanta*, American Society for Metals.
Ohmae, N., Nakai, T., and Tzukizoe, T. (1974a). *Wear* **30**, 299.
Ohmae, N., Nakai, T., and Tzukizoe, T. (1974b). *Japan. J. Appl. Phys.*, Suppl 2, Pt 1, p. 451.
Ohmae, N. (1976). *J. Vac. Sci. Technol.* **13**, 82.
Ohsaki, T., Yoshida, M., Fukube, Y., and Nakamura, K. (1977). *Thin Solid Films* **45**, 563.
Parent, E. D. (1978). *Proc. 21st Ann. Conf. Soc. Vacuum Coaters, Detroit, Michigan*, p. 8.

Parker, P. (1950). *Electronics*, Edward Arnold Ltd, London.
Patten, J. W. (1978). *Thin Solid Films* **54**, 325.
Pawlewiez, W. T., and Loegreid, N. (1978). *Proc. SPIE* **140**, 156.
Pawlewiez, W. T., Martin, P. M., Hays, D. D., and Mann, I. B. (1982). *Proc. SPIE* **325**, 105.
Perry, A. J. (1983). *Thin Solid Films* **107**, 167.
Perry, A. J., and Archer, N. J. (1980). *AGARD Lecture Series* No. 106, *Materials Coating Techniques*.
Plaisted, A. (1978). *Proc. 21st Ann. Conf. Soc. Vacuum Coaters, Detroit, Michigan*, p. 77.
Pulker, H. K. (1976). *Reactive Ion Plating, 2nd Coll. Int. Pulverisation Cathodique et ses Applic.*, CIP 76, Nice, France, p. 105.
Pulker, H. K. (1984). *Coatings on Glass*, Elsevier Scientific Publishers, Amsterdam.
Pulker, H. K. Paesold, G., and Ritter, E. (1976). *Appl. Optics* **15**, 2986.
Reale, G. (1978). *Bulg. Akad. Na Nawk* **31**, 281.
Reid, I. M., Macleod, H. A., Henderson, E., and Carter, M. J. (1979). *Proc. Conf. Ion Plating and Allied Techniques, IPAT 79*, CEP Consultants, Edinburgh, p. 114.
Report (1984). *The Military Critical Technologies List*, Report No. FTC 6346/1/2, US Department of Defence, October 1984.
Report (1985). *Plating and Surface Finishing*, P & SF Report, p. 20, July 1985.
Restall, J. E. (1979). 'Improvements in or relating to nickel-chrome-base alloys', Brit. Patent Appl. No. 7925846.
Restall, J. E. (1981). 'Surface degradation and protective treatments' in *The Development of Gas Turbine Materials*, Ed. G. W. Meetham, Applied Science Publishers, London, Ch. 10.
Reynolds, L. F. (1985). 'Evaluation of new surface coatings for corrosion protection of unpeened springs', *Report* No. 381, Spring Research Manufacturers Association (SRAMA), Sheffield, UK.
Ridge, M. I., Stenlake, M., Howson, R. P., and Bishop, C. A. (1981). *Thin Solid Films* **80**, 31.
Ridge, M. I., Howson, R. P., and Bishop, C. A. (1982). *Proc. SPIE, Technical Meeting, Los Angeles*, p. 325.
Ritter, E. (1981). *Appl. Optics* **20**, 21.
Salama, M. S. (1983). *Proc. Int. Ion Engineering Congress, ISIAT 83 and IPAT 83*, Kyoto, Japan, p. 871.
Salem, F. B. (1977). *PhD Thesis*, University of Salford, UK.
Salem, J., and Sequeda, F. (1981). *J. Vac. Sci. Technol.* **18**, 149.
Sanders, J. V. (1971). *Chemisorption and Reactions on Metallic Films*, Ed. J. R. Anderson, Academic Press, London.
Sato, T., Tada, M., Huang, X. C., and Takei, H. (1978). *Thin Solid Films* **54**, 61.
Schiller, S., Beister, G., Schneider, S., and Wieber, W. (1980). *Thin Solid Films* **72**, 475.
Schiller, S., Heisig, U., and Goedicke, K. (1977). *Proc. 7th Int. Vac. Conf. and 3rd Int. Congr. Solid Surfaces, Vienna*, p. 1545.
Schiller, S., Heisig, U., Steinfelder, K., and Strumpter, J. (1979). *Thin Solid Films* **64**, 455.
Schiller, S., Lenk, P., Foerster, H., Jaesche, G., and Kuehn, G. (1974). *Proc. 3rd Electron Beam Processing Seminar, Stratford-upon-Avon, UK*. Published by the University Technology Corporation, Dayton, Ohio, p. 57.
Schorer, M., and Wirz, P. (1984). *Thin Solid Films* **119**, 203.
Sherbiney, M., (1975) *PhD Thesis*, University of Salford, UK.
Sherbiney, M., and Aboukhashaba, A. (1983). *Proc. Ion Engineering Congress, ISIAT 83 and IPAT 83, Kyoto, Japan*, p. 1275.
Sherbiney, M., and Halling, J. (1977). *Wear* **45**, 211.
Sigmund, P. (1969). *Phys. Rev.* **184**, 383.
Sims, R. I. (1976). *Metallurgist Mats Technologist* **10**, 21.
Smith, H. R. (1974). *Proc. 17th Ann. Conf. Soc. Vacuum Coaters*, p. 54.
Spalvins, T. (1970). 'Deposition of alloy films on complex surfaces by ion plating with flash evaporation', NASA TMX-52823.
Spalvins, T. (1971). *Lubrication Engineering* **27**, 40.
Spalvins, T. (1974). *Metal Finishing* **12**, 38.
Spalvins, T. (1980). *J. Vac. Sci. Technol.* **17**, No. 315.
Spalvins, T. (1981). *Thin Solid Films* **84**, 267.
Spalvins, T. (1985). 'The structure of ion plated films in relation to coating properties'. Paper presented at *ASM Tech. Div Conf, Atlanta, Georgia*.
Spalvins, T. and Brainard, W. A. (1976). 'Ion plating with an induction heating source', NASA TMX-3330.

Spalvins, T., Przybyszewski, J. S., and Buckly, D. H. (1966). *NASA Tech. Note* TND-3707.
Spencer Franklin Ivadizer Unit (1985–1986). Leeds, UK. Private Communication.
Sproul, W. D. (1984). *Thin Solid Films* **118**, 279.
Sproul, W. D. (1985). *J. Vac. Sci. Technol.* **A3**, 580.
Sproul, W. D. (1986). *Proc. Int. Conf. Thin Films, ICTF6*, to be published in *Thin Solid Films.*
Steffans, H. D. (1980). Paper 5, AGARD-65-106, NATO.
Stern, E., and Coswell, H. L. (1967). *J. Vac. Sci. Technol.* **4**, 128.
Stockbridge, C. D. (1966). *Vacuum Microbalance Techniques*, Ed. K. H. Behrndt, Plenum Press, New York, p. 193.
Sundgren, J. E. Johansson, B. O., and Karlsson, S. E. (1983). *Thin Solid Films* **105**, 353.
Sundquist, H. A., Sirvio, E. H., and Kurkinen, M. T. (1982). *Proc. Conf. Ion Assisted Surface Treatments, Techniques and Processes, Warwick University, Uk*, p. 7.1. Published by the Metals Society, UK.
Suzuki, K., and Howson, R. P. (1983). *Proc. Int. Ion Engineering Congress, ISIAT 83 and IPAT 83, Kyoto, Japan*, p. 889.
Suzuki, K., Matsumoto, K., and Takatsuka, T. (1981). *Thin Solid Films* **80**, 67.
Swaroop, B., Meyer, D. E., and White, G. W. (1976). *J. Vac. Sci. Technol* **13**, 680.
Takagi, T., Yamada, I., and Sasaki, A. (1976). *Thin Solid Films* **39**, 207.
Takagi, T. (1982). *Thin Solid Films* **92**, 1.
Takagi, T. (1986). *Vacuum* **36**, 27.
Takai, O. (1983). *Proc. Int. Ion Engineering Congress, ISIAT 83 and IPAT 83, Kyoto, Japan*, p. 865.
Talboon, F. P., Elam, R. C., and Wilson, L. W. (1969). NASA CR72813.
Taylor, E. (1969). *SPS Report* No. 1889.
Technology Forecast (1986). Published in *Metal Progress*, January, p. 45.
Teer, D. G. (1975). *Tribology International* **8**, 247.
Teer, D. G. (1977). *J. Adhesion* **8**, 289.
Teer, D. G. (1983). *Coating for High Temperature Applications*, Ed. E. Lang, Applied Science Publishers, London, p. 83.
Teer, D. G., and Arnell, R. D. (1974). UK Patent Application No. 29701174.
Teer, D. G., Delcea, B. L., and Kirkham, A. J. (1976). *J. Adhesion* **8**, 171.
Teer, D. G., and Halling, J. H. (1978). *Advances in Surface Coating Technology; Int. Conf. Welding Inst, London*, p. 123.
Theard, L. P. (1968). *Rev. Sci. Instrum.* **39**, 1639.
Thornton, H. R., and Wolanski, Z. R. (1967). *Lab. Eng.* **22**, 271.
Thornton, J. A. (1974). *J. Vac. Sci. Technol.* **11**, 660.
Thornton, J. A. (1977). *Ann. Rev. Material Science* **7**, 239.
Thornton, J. A. (1978). *J. Vac. Sci. Technol.* **15**, 171.
Thornton, J. A. (1979). *Metal Finishing*, p. 83.
Thornton, J. A., and Hoffman, D. W. (1981). *J. Vac. Sci. Technol.* **18**, 203.
TiN Equipment Suppliers: Ulvac, Tokyo, Japan. TECVAC, Cambridge, England. Leybold-Heraeus GmbH, Hanau, West Germany. Multi-Arc Vacuum Systems Inc, USA.
Ting, L. Y. (1982). *J. Vac. Sci. Technol.* **21**, 14.
Todd, M. J., and Bentall, R. H. (1978). *Proc. 2nd conf. solid lubrication, Denver, 1978, Pub. ASLE Spec. Pub.* **6**, p. 148.
Tolfree, D. W. L., Teer, D. G., and Salama, M. (1979). *Proc. Conf. Ion Plating and Allied Techniques*, CEP Consultants, Edinburgh, p. 144.
Trent, E. M. (1979). *Treatise in Materials Science and Technol.* **13**, 443.
Tsaur, B. V., Hatteson, S., Chapman, G., Lian, Z. L., and Nicolet, M. A. (1979). *Appl. Phys. Lett.* **35**, 825.
Uetz, H., and Fohl, J. (1979). *Wear* **49**, 253.
Valizadeh, H. (1982). *PhD Thesis*, University of Salford, UK.
Valli, J. (1984). Paper presented at the ICTF-6 Conf, Stockholm.
Varasi, M., Misiano, C., Mancini, C., and Sartori, P. (1986). *Vacuum* **36**, 143.
Vincett, P. S., Barlow, W. A., and Robert, G. G. (1977). *J. Appl. Phys.* **48**, 3800.
Vossen, J. L., O'Neill, J. J., Mesker, O. R., and James, E. A. (1977). *J. Vac. Sci. Technol.* **14**, 85.
Vossen, J. L., and Kern, W. (Eds) (1978). *Thin Film Processes*, Academic Press, New York.
Walls, J. M., Hall, D. D., Teer, D. G., and Delcea, B. L. (1978). *Thin Solid Films* **54**, 303.
Wan, C. T., Chambers, D. L., and Carmichael, D. C. (1971). *J. Vac. Sci. Technol.* **8**, VM99.
Wan, C. T., Chambers, D. L., and Carmichael, D. C. (1973). *Proc. 4th Int. Conf. Vacuum Metallurgy, Tokyo*, p. 150.

Weissmantel, C. (1976). *Thin Solid Films* **32**, 11.
Weissmantel, C. (1977). *Proc. 7th Int. Vac. Congr. and 3rd Int. Conf. Solid Surfaces, Vienna*, p. 1533.
Weissmantel, C., Fiedler, O., Hecht, G., and Reisse, G. (1976). *Thin Solid Films* **13**, 359.
Weston, G. F. (1984). *Phys. Technol.* **15**, 37.
White, G. W. (1973). *Research/Development* **24**, 43.
White, G. W. (1974). 'A high rate source for ion plating', *Vacuum Metallurgy Conf.*, AVS-VAC, Metall. Div., Pittsburgh, PA. p. 98.
White, M. S. (1978). *Product Finishing*, p. 4.
Wick, C., and Velleux, R. F. (Eds) (1985). *Tool and Manufacturing Engineering Handbook*, 4th edn, Ch. 16, pp. 16–83, published by the Society of Manufacturing Engineers, Michigan.
Wickersham, C. E., and Pool, J. E. (1985). *J. Vac. Sci. Technol.* **A3**, 581.
Wieder, H. H. (1970). *Intermetallic Semiconducting Films*, Pergamon Press, New York.
Winters, H. F., and Kay, E. (1967). *J. Appl. Phys.* **38**, 3928.
Winters, H. F., and Kay, E. (1972). *J. Appl. Phys.* **43**, 794.
Wisander, D. (1971). *NASA Tech. Note* 6455.
Wittmer, M., Studer, B., and Melchier, H. (1981). *J. Appl. Phys.* **52**, 5722.
Wright, P. K. (1981). 'Wear of high speed steel tools', Paper presented at a Fagersta-sponsored Symposium, Sweden.
Yamada, I., and Takagi, T. (1981). *Thin Solid Films* **80**, 105.
Yarnold, G. D., and Bolton, H. C. (1949). *J. Sci. Instrum.* **26**, 38.
Young, R. W., and Hurley, R. E. (1979). *Proc. Conf. Ion Plating and Allied Techniques, IPAT 79*, CEP Consultants, Edinburgh, p. 136.
Zega, B., Kornmann, M., and Amignet, J. (1977). *Thin Solid Films* **54**, 577.
Zum Gahr, K. H. (1979). *Metal Progress* **116**, 461.

Index